Geothermal Energy Research Trends

GEOTHERMAL ENERGY RESEARCH TRENDS

HERMAN I. UECKERMANN
EDITOR

Nova Science Publishers, Inc.
New York

For permission to use material from this book please contact us:
Telephone 631-231-7269; Fax 631-231-8175
Web Site: http://www.novapublishers.com

LIBRARY OF CONGRESS CATALOGING-IN-PUBLICATION DATA

Geothermal energy research trends / Herman I. Ueckermann, editor.
p. cm.
Includes bibliographical references and index.
ISBN-13: 978-1-60021-683-1 (hardcover)
ISBN-10: 1-60021-683-8 (hardcover)
1. Geothermal engineering. 2. Geothermal resources. I. Ueckermann, Hermann I.
TJ280.7.G4425 2007
621.44--dc22

2007012188

Published by Nova Science Publishers, Inc. ✦ New York

CONTENTS

PREFACE

Geothermal energy is a form of renewable energy derived from heat deep in the earth's crust. Enormous amounts of thermal energy are continuously generated by the decay of radioactive isotopes of underground rocks and stored in our globe's interior. This heat is as inexhaustible and renewable as solar energy. The heat is brought to the near-surface by thermal conduction and by intrusion into the earth's crust of molten magma originating from great depth. As groundwater is heated, geothermal energy is produced in the form of hot water and steam. The heated groundwater can be used for direct heating of homes and greenhouses, for vegetable drying, and for a number of other uses. These are known as direct uses of geothermal energy. Geothermal energy is also used for electricity production. This new book presents leading-edge research in a field destined for increased attention throughout the world.

Temperature profiles have been empirically investigated in the underground geological formations of the Nea Kessani (Greece) geothermal system by the Greek Institute of Geology and Mineral Exploration using measurements in a set of vertical drill holes. In Chapter 1, the authors used the BME method to derive spatial temperature estimates in the Nea Kessani region in a mathematically rigorous and scientifically meaningful manner. The proposed analysis involves the solution of a stochastic partial differential equation representing the geothermal field and is conditioned on site-specific information (random boundary conditions reflecting in-situ uncertainties etc.). Temperature probability distributions were generated at the nodes of a dense spatial grid, which can provide a detailed understanding of the geothermal situation by means of various temperature maps (most probable, error minimizing etc.), depending on the objectives of the study. The BME solutions are more informative than the direct (analytical and numerical) solutions of the geothermal model obtained in a formal mathematical sense.

Chapter 2 introduces briefly distribution of hydrothermal resources, potential of hydrothermal energy, geochemistry of geothermal fluids and correlation between geothermal areas and seismic zones in China. More than 3,200 hydrothermal manifestations have been found in China. About 2,240 drilled wells reveal that 275 high temperature sites of hydrothermal energy, which are expected to supply a need of electric generators with total annual output of 5,800 MW. More than 2,900 sites of low and intermediate temperature geothermal systems have been found, which can be utilized for heating, medicine treating, bathing, farming, etc.

Most geothermal waters in China are (Na, Ca)-HCO_3 type, and some are (Ca, Na)-SO_4 and Na-Cl types. Stable isotopic compositions of oxygen and hydrogen indicate the geothermal waters are derived from meteoric water, with small amount of magmatic volatile. Reservoir temperatures calculated with chemical geothermometers range from about 100 °C to 350 °C. Geochemical variations of geothermal fluids with time are found, which are correlated to hydrothermal eruption, earthquakes and exploitation.

Main gaseous components of geothermal systems in China are CO_2, N_2, O_2, and trace amount of H_2S, H_2, CH_4, NH_3, CO, C_2H_6, C_3H_8 as well as noble gases (Rn, He, Ar, Ne, Kr, Xe). The gaseous concentrations of geothermal systems are correlated to the temperatures of geothermal systems and seismic faults. The gases have a multiple origins of crust, mantle and atmosphere.

The chapter emphasizes both the spatial correlation between the geothermal areas/zones and the seismic zones and the energetic relationship between geothermal-fluid geochemistry and seismic activity. The larger amounts of mantle gases the geothermal systems contain, the higher temperature of geothermal systems and the more active the seismic zones. The deep earth fluids provide both matter and energy for geothermal fields and earthquake generation, and carry the messages of geothermal reservoir and earthquake.

As presented in Chapter 3, the Larderello and Mt. Amiata geothermal fields in Tuscany are large active thermal systems. Both likely overlie young plutonic rocks that serve as the principal sources of heat. The features of the two geothermal systems are similar. *1) Structural setting.* The geothermal fields of Larderello and Mt. Amiata are located in the inner part of the Northern Apennines, characterized asthenosphere uplift and delamination of the crustal lithosphere or underplating. *2) The heat source* both at Larderello and Mt. Amiata can be ascribed to the presence of shallow igneous intrusions. *3) The heat flow data* for the area surrounding both the Larderello and Mt. Amiata geothermal fields show a comparable areal extension and similar values (up to 200-300 mW/m^2). *4) Cap rocks and reservoirs.* Both the Larderello and the Mt. Amiata fields have shallow vapor-dominated sedimentary, and deep metamorphic reservoirs. At Larderello super-heated steam is present in both reservoirs, to depth of more than 3.5 km, whereas the deep reservoir of the Mt. Amiata geothermal fields is likely water-dominated. In both fields the upper reservoir is present below the flysch units forming the cap rocks. *5) Permeability* is due to rock fracturing, even at depths of about 4 km and temperatures as high as 350°C. Pressure greater than hydrostatic and a supercritical fluid can occur in the deepest part of the geothermal fields. *6) Hydrothermal alteration and contact metamorphism.* At Larderello and Mt. Amiata there is evidence of an early contact metamorphism related with the intrusion of the granites. *7) Recharge.* The water stable isotope values of the steam discharged by the geothermal wells at Larderello indicate a meteoric origin. A geochemical regional study on the thermal waters and gases of the Mt. Amiata area indicates that the geothermal reservoirs originated from a meteoric fluid, mainly stored in a regional Mesozoic dolomite-anhydrite unit, and evolved in a Na-Cl, CO_2 gas-reach reservoir by interaction with calcite-bearing metamorphic rocks.

The high temperatures existing in correspondence of a deep seismic reflector suggest the occurrence of a deep-seated unconventional geothermal resource (UGR), which can be possibly exploited. The heat could be mined from silica-rich rocks close to a plastic state, but where fracturing can be induced by fluid overpressure and abrupt high strain rates. This geothermal resource is very important, and requires a re-assessment of the geothermal resources in Italy, considering the possibility of the exploitation of the new reservoirs.

High-temperature geothermal reservoir is under consideration, consisting of two high-permeability layers, which are separated by a low-permeability stratum. The thermodynamic conditions are assumed to imply that the upper and lower high permeability layers are filled in by water or by vapor, respectively. The stable stationary regimes of vertical phase flow between water and vapor layers in the low permeability stratum may exist. We give possible types of transition to instability of the vertical flows in such a system under the condition of smallness of the advective heat transfer in comparison with the conductive one. It is found that in the generic case there exist three different scenarios of the instability onset of the stationary vertical flows. They are accompanied by the bifurcations of solutions describing the destabilizing vertical flows. The possible scenarios of the evolution of the system over the threshold of instability are discussed in Chapter 4.

As explained in Chapter 5, abundant genetic and sedimentary indicators have been found in the thick coal beds from three fault-controlled coal basins on the Central Massif, France. A new formation model for thick continental (intra-mountainous) lacustrine peat swap is proposed. In the new coal accumulation mechanism, thick coal beds were associated with various gravity-influenced breccia and sandstone interlayer sediments and the subaquatic gravitary current transported the organic (peat) and inorganic clasts formed in lakeshore swamp were formed in active clastic environment, and were associated with various gravity-influenced mudstone and sandstone interlayers. The presence of a great number of gravity-flow sediments such as detrital flow, diluted slurry flow or turbidity-current sediments in the coal seams, and that of the contemporaneous gravity slump and deformation structures in the coal seams both indicate that the accumulation of the thick coal beds was characterized by the relatively deep water environment and allochthonous sedimentation. This new model interprets reasonably the accumulation mechanism of the thick coal beds developed in the fault basins in the Central Massif (France) and provides a completely new idea with respect to the traditional coal formation models.

As an innovation, the authors propose that a new frontier in geothermal research should be explored that involves the coupling of thermal and chemical simulations in an integrated "magma chamber-reservoir" model. To achieve this innovation in geothermal research, the authors have written a new computer program (in Fortran 90) in modular structure that runs on a PC under the dynamic memory concept and simulates heat transfer conductive and convective processes both in a magma chamber and the overlying geothermal reservoir as well as computes *in-situ* major element chemistry of magmas that evolved in the magma chamber as a result of processes of assimilation, crystallization (liquid line of descent), magma mixing, recharge, and eruption. This combined task is accomplished in three dimensions (3-D) – a substantial improvement as compared to the current practice of obtaining thermal solutions in 1-D or 2-D and of modeling chemical data obtained from the analysis of surface outcrops *without* reference to the actual location within the Earth where the magmas were stored prior to eruption. In fact, if temperature estimates in drill wells and chemical data for surface rocks were available, this information could be used to constrain the model we are proposing as a new research frontier. The practice of "direct" modeling can be replaced in the future by inverse modeling when greater computing and storage capacities of personal computers will be available. Chapter 6 briefly reviews the current state of thermal modeling of geothermal areas and presents the salient features of the authors' new research approach, including a brief description of their computer program. An application example of a Mexican geothermal field (Los Humeros, Puebla), currently under exploitation for electricity production, will highlight the use of their software. This particular geothermal field was chosen for illustration purposes because of the availability of required thermal and chemical data to test the 3-D simulation model. The authors have successfully reproduced some of the major element chemical characteristics for the most voluminous caldera-forming eruption at about 0.46 Ma and the present-day thermal regime inferred from static formation temperatures using a quadratic regression of the actually measured bottom hole temperature data.

The inflow curves are equivalent to the characteristic curves of production at bottom-hole conditions. They have been used for several decades as a tool for the analysis of the productivities of wells, to establish criteria for their operation. The development and application of the inflow curves begins in petroleum engineering and later on with adaptations, according to the type of fluid, they are applied to geothermal engineering.

In Chapter 7 the equations of inflow behavior along with damage effect for oil systems are shown. The existent relationships of the inflow behavior in geothermal reservoir engineering are revised. The relationships of the inflow behavior of geothermal fluid are exposed and discussed, considering the fluid as a ternary mixture (H_2O-CO_2-$NaCl$) for low and high salinity. Those factors influencing the presence of damage are described and the parameters applicable to geothermal systems are thus obtained. Representative variables of the damage effect and the average physical characteristics of a geothermal system, within the relationships of the inflow behavior, are incorporated. The acquisition of the first geothermal inflow type-curve for different values of the damage is shown.

The geothermal inflow type-curves obtained for different damage values are applicable to data from production tests in geothermal wells. They are a useful tool in determining the value of the damage in the well. The methodology implies the selection of the corresponding geothermal type-curve according to the characteristics of fluid salinity.

The methodology proposed to determine the value of the damage effect in a well is described, starting from the geothermal inflow type-curves using data from a production test. The authors incorporate the method of Jones *et al.* (1976) as a verification tool for the presence or absence of damage.

The authors show the use of this methodology with data from production tests in wells from three different fields, using the two geothermal inflow type-curves (for low and high salinity). The ascertained damage values are verified in the results of Jones' method, finding consistency in both techniques.

This methodology is useful as a tool for the diagnosis of the damage effect starting from production tests of a geothermal well. Identifying the numerical value of the damage permits the establishment of criteria for preventive and remedial operations in such wells.

RESEARCH AND REVIEW STUDIES

In: Geothermal Energy Research Trends
Editor: Herman I. Ueckermann, pp. 3-17
ISBN: 1-60021-683-8

Chapter 1

BME-GENERATED TEMPERATURE MAPS OF THE NEA KESSANI GEOTHERMAL FIELD

Konstantinos Modis[a], Hwa-Lung Yu[b], George Christakos[b], Robert Stewart[c] and George Papantonopoulos[a]

[1a] School of Mining and Metallurgical Engineering, National Technical University of Athens, Greece,
[b]Department of Geography, San Diego State University
San Diego, California, USA.
[c]The Institute of Environmental Modelling, University of Tennessee
Knoxville, Tennessee, USA

Abstract

Temperature profiles have been empirically investigated in the underground geological formations of the Nea Kessani (Greece) geothermal system by the Greek Institute of Geology and Mineral Exploration using measurements in a set of vertical drill holes. In this work, we used the BME method to derive spatial temperature estimates in the Nea Kessani region in a mathematically rigorous and scientifically meaningful manner. The proposed analysis involves the solution of a stochastic partial differential equation representing the geothermal field and is conditioned on site-specific information (random boundary conditions reflecting in-situ uncertainties etc.). Temperature probability distributions were generated at the nodes of a dense spatial grid, which can provide a detailed understanding of the geothermal situation by means of various temperature maps (most probable, error minimizing etc.), depending on the objectives of the study. The BME solutions are more informative than the direct (analytical and numerical) solutions of the geothermal model obtained in a formal mathematical sense.

Keywords: Geothermal, geophysics, stochastic, modelling, BME.

[1] E-mail address: kmodis@mail.ntua.gr. To whom correspondence should be addressed.

1. Introduction

A composite solution of a physical partial differential equation (PDE) representing a geophysical system not only assures consistency of the derived solution with the physical PDE but it also accounts for the multi-sourced uncertainty that characterizes the real-world system and incorporates important site-specific information available about the system (Christakos, 2004, 2005). The composite solution is distinct from the standard PDE solution in a direct (formal) sense that isolates the geophysical system from the effects of site-specific effects and extracts from mathematics an attribute-value at each specified point so that the PDE is satisfied (e.g., Zwillinger, 1989).

The BME approach was first introduced by Christakos (1990, 1991) and later developed in a more general epistematics context. In earth sciences, the BME approach has been used to derive composite solutions of various physical PDE (see literature review in Yu *et al.*, 2006). In this work, we generate informative temperature maps in terms of the composite BME solution of a stochastic three-dimensional geothermal PDE under random boundary conditions (BC) in the Nea Kessani region (Greece). In addition, the solution derived by BME will account for the site-specific data available from a drilling survey conducted by the Greek Institute of Geology and Mineral Exploration (IGME) during the period 1980-1991 (Kolios, 1993). The thus obtained solution will be compared with the analytical solution of the same geothermal PDE subject to specific BC and the direct (numerical) solution of PDE subject to random BC.

2. The BME Approach of Epistematics

Let $T(\boldsymbol{s})$ be the temperature random field (RF; Christakos, 1992), where $\boldsymbol{s}=(s_1,s_2,s_3)$ are three-dimensional coordinates. The BME distinguishes between two major knowledge bases (KB): the general (or core) KB, indicated by *G*, which includes mathematical models, physical laws and scientific theories; and the site-specific (or specificatory) KB, indicated by *S*, which assimilates different information sources about the particular situation of interest--these sources include hard (exact) measurements, soft (uncertain) data and empirical relations. The composite BME solution of a geothermal PDE has certain important properties (for details, see Christakos, 2000, 2005): it considers the geothermal PDE as an incomplete representation of reality, in which the auxiliary model parameters and outcome field have significant uncertainty features that need to be accounted for; there may exist a wealth of knowledge apart from the physical model that is directly related to it and also needs to be taken into consideration. As a result, a composite temperature solution maximizes the information provided by the geothermal model (teleologic ideal) and, at the same time, it assimilates all relevant knowledge sources (evolutionary adaptation ideal). In mathematical terms, the BME approach replaces the PDE with a set of fundamental equations

$$\left.\begin{aligned} &\int d\tau(\boldsymbol{g}-\overline{\boldsymbol{g}})e^{\boldsymbol{\mu}^T\boldsymbol{g}}=0 \\ &\int d\tau\,\xi_S\, e^{\boldsymbol{\mu}^T\boldsymbol{g}}-A\,f_K(\boldsymbol{s})=0 \end{aligned}\right\}, \tag{1}$$

where τ denote temperature $T(\boldsymbol{s})$-realizations across space; $\boldsymbol{g}$ is a vector of g_α-functions ($\alpha = 1,...,N$) that represents stochastically the physical model under consideration as well as other *G-KB*, if available (the bar denotes stochastic expectation); μ is a vector of μ_α-coefficients that depends on the spatial coordinates and is associated with $\boldsymbol{g}$ (i.e., the μ_α express the relative significance of each g_α-function in the composite solution); the ξ_S represents the *S*-KB available; A is a normalization parameter; and f_K is the probability density function (pdf) of the composite solution at each point (the subscript $K = G \cup S$ means that the pdf integrated the *G*- and *S*-KB). The unknowns in Eqs. (1) are the μ and f_K across space. Given the pdf f_K, different temperature values can be derived at each solution node (most probable, error minimizing etc.), depending on the objectives of the study.

3. Stochastic Formulation of the KB about the Geothermal Field

3.1. Geology and Geophysics of the Region

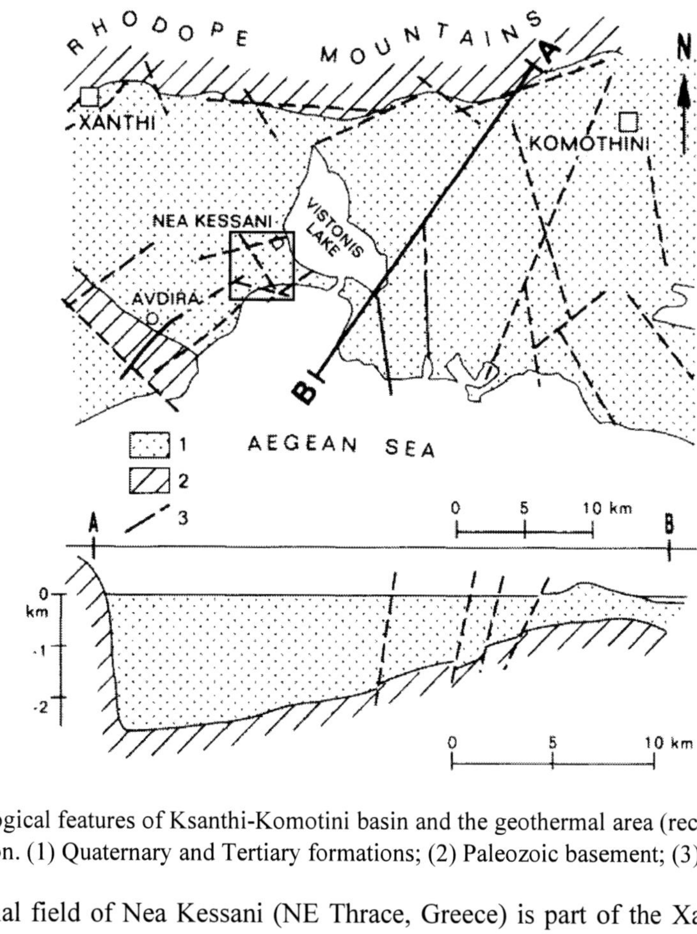

Figure 1. Geological features of Ksanthi-Komotini basin and the geothermal area (rectangle) along with a vertical section. (1) Quaternary and Tertiary formations; (2) Paleozoic basement; (3) fault.

The geothermal field of Nea Kessani (NE Thrace, Greece) is part of the Xanthi - Komotini (post-orogenic Tertiary sedimentary) basin, which is located next to the Aegean Sea and the Vistonis Lake. The field covers an area of about 1600 km^2 between the Rhodope Mountains and the Aegean coast (Fig. 1). The basin is constituted mainly of clastic sediments. It reaches its maximum depth at the foot of the Rhodope chain, and its minimum depth in the vicinity of the coast, where the Nea Kessani geothermal field of interest is located. The geothermal field is created by heat distribution in the geological formations of the ground at some depth due to thermal fluid circulation (Kolios, 1993). This causes an increase of the earth natural geothermal gradient (which is 30 °C/km on average) by several degrees reaching 35°C/km. The hot reservoir created by the thermal fluids has an average temperature of 75 – 80°C and covers an area of 5 km^2. The reservoir roof is 100-120m deep (southern part) near Aegean Sea; its basement in the same area is 450m deep. At the north part of the reservoir, its roof is 300-350m deep and its basement is located at a depth greater than 1 km. A detailed description of the region with relevant maps is found in Papantonopoulos and Modis (2005) and references therein.

3.2. The S-KB

In order to study the geothermal anomaly in the region, twenty five (25) exploratory boreholes were drilled by the Greek IGME during the time period 1980-1991: (*a*) the first phase of the drilling campaign began in 1980 and produced 11 drill holes in depths varying from 65 to 475 m; (*b*) the second phase, which started in 1990, produced 14 drill holes to depths varying from 200 to 500 m. The drill-hole datasets consist of temperature measurements taken within the drill wells by an electrical resistance thermometer with ± 0.1°C measurement error. In Fig. 2 the locations of the drill-hole collars are shown along with two vertical sections showing the temperature distribution in the area. The drill-hole data used in this work are given in Table 1.

Table 1. Drill-hole collar coordinates and other data.

Drill hole	s_1-coordinate (m)	s_2-coordinate (m)	Max temperature measured (°C)	Depth of maximum temperature (m)
G1	1150.88	645.61	50.50	390
G2	364.91	56.14	39.40	294
G3	280.70	1319.30	83.00	460
G5	0	2245	41.60	370
G7	1010.53	1992.98	61.90	260
G8	645.61	1656.14	70.40	226
G9	2189.47	1880.70	70.00	272
G12	1656.14	1880.70	73.70	350
G13	898.25	1908.77	79.00	350
G15	56.14	898.25	67.20	160
G17	28.07	1403.51	74.50	170
G18	1543.86	870.18	41.60	288
G21	421.05	1768.42	71.70	340
G23	2245.61	0.00	39.80	390
G24	2105.26	1178.95	71.90	420
G25	2245.61	2245.61	51.20	316

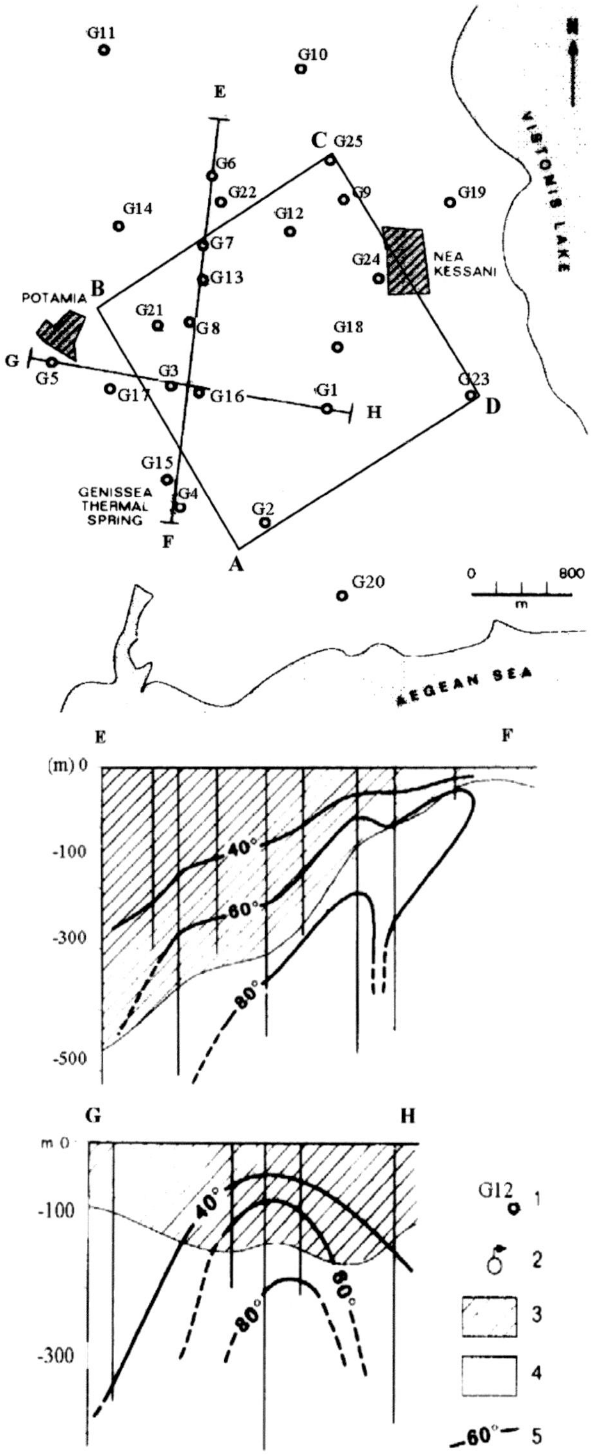

Figure 2. Drill hole location and vertical cross-sections in the area of interest (dashed parallelogram). In sections AB and CD the temperature distribution derived from the drill data is presented. (1) Drill collar (2) Thermal spring (3) roof formations (4) reservoir (5) isotherms in °C.

In view of the above, the *S*-KB includes precise down-the-hole temperature measurements (hard data set) obtained from the drilling survey. The domain considered in the Nea Kessani geothermal study is an orthogonal parallelepiped located 100m below the surface and having dimensions: 2250m (width) × 2250m (length) × 300m (height). The *S*-KB is stochastically formulated in terms of the functions $\boldsymbol{\xi}_S$ of Eq. (1), i.e.

$$\left.\begin{array}{l}\text{Hard data}: T(\boldsymbol{s}) \text{ measurements } (90) \\ \text{Soft data}: \quad T(\boldsymbol{s}) \text{ ranges } (49)\end{array}\right\} \rightarrow \boldsymbol{\xi}_S. \tag{2}$$

Explicit probabilistic expressions of $\boldsymbol{\xi}_S$ can be found in the literature (Christakos, 2000; Kolovos *et al.*, 2002). The locations of the hard data available are shown in Fig. 3.

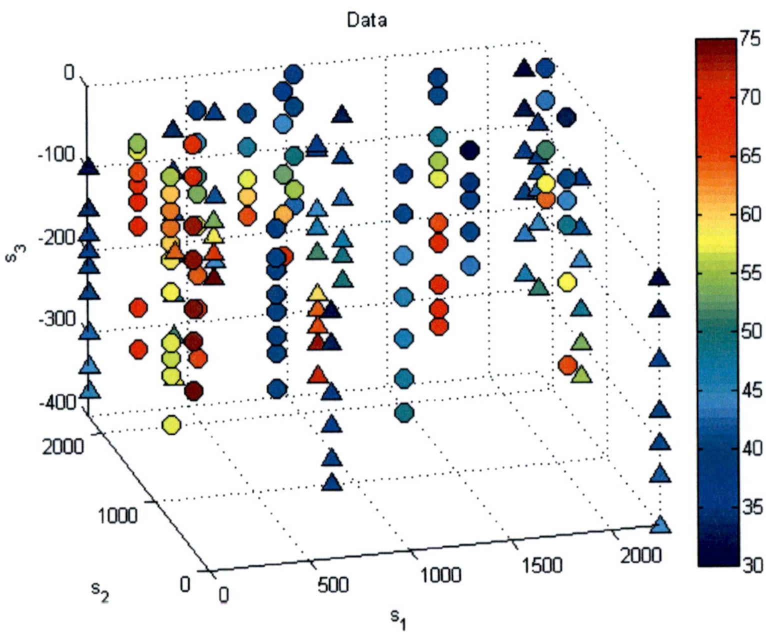

Figure 3. The three-dimensional plot of the observations (in °C)at drill-holes where the circles are the hard data and triangles are the interval (uncertain) data.

3.3. The G-KB

Assuming an isotropic three-dimensional region of uniform thermal conductivity with no heat generation, the heat transfer law in the Nea Kessani study is core knowledge (*G*-KB) expressed as follows

$$\nabla^2 T(\boldsymbol{s}) = 0, \tag{3}$$

where $T(\boldsymbol{s})$ is the temperature RF and $\boldsymbol{s} = (s_1, s_2, s_3)$. The corresponding boundary conditions (BC) are given by

$$T(\mathbf{s}) = A(\mathbf{s}) + \varepsilon_A(\mathbf{s}), \quad T(\mathbf{s}) \in \Gamma_D. \tag{4}$$

where the Γ_D denotes that the stochastic Dirichlet BC are considered; the $A(\mathbf{s})$ is a deterministic temperature trend and the random fluctuation $\varepsilon_A(\mathbf{s})$ expresses uncertainty in our knowledge of the BC (other forms of BC are discussed in Yu *et al.*, 2006). We do not merely seek the direct mathematical solution of Eq. (3), isolated from other influences. Instead, we seek composite solutions of the geothermal situation that are physically consistent with (3) in a stochastic sense that accounts for the uncertainty expressed by Eqs. (4) and, in addition, they incorporate empirically important site-specific information sources.

The temperature RF consists of a deterministic spatial mean $\overline{T}(\mathbf{s})$ and a fluctuation field $\tilde{T}(\mathbf{s})$, i.e. $T(\mathbf{s}) = \overline{T}(\mathbf{s}) + \tilde{T}(\mathbf{s})$. Then, the heat transfer law (2) can be decomposed into a deterministic part

$$\nabla^2 \overline{T}(\mathbf{s}) = 0, \tag{5}$$

with BC

$$\overline{T}(\mathbf{s}) = A(\mathbf{s}), \quad \overline{T}(\mathbf{s}) \in \Gamma_D; \tag{6}$$

and a random part

$$\nabla^2 \tilde{T}(\mathbf{s}) = 0, \tag{7}$$

with BC

$$\tilde{T}(\mathbf{s}) = \varepsilon_A(\mathbf{s}), \quad \tilde{T}(\mathbf{s}) \in \Gamma_D. \tag{8}$$

Because of the assumption of homogeneous-isotropic thermal conductivity, the uncertainty associated with the temperature field in the above equations is due to the BC uncertainty.

The covariance function of $\tilde{T}(\mathbf{s})$, denoted by c_T, is derived from Eq. (7) if one multiplies $\tilde{T}(\mathbf{s})$ by $\tilde{T}(\mathbf{s}')$, $\mathbf{s}' = (s_1', s_2', s_3')$, and take the stochastic expectation of the product leading to the Laplacian

$$\nabla^2 c_T(\mathbf{s}, \mathbf{s}') = 0, \tag{9}$$

with BC

$$c_T(\mathbf{s}, \mathbf{s}') = \overline{\varepsilon_A(\mathbf{s}) \tilde{T}(\mathbf{s}')}, \quad c_T(\mathbf{s}, \mathbf{s}') \in \Gamma_D. \tag{10}$$

For each pair of grid nodes $(\boldsymbol{s},\boldsymbol{s}')$, Eqs. (9)-(10) should be solved for $c_T(\boldsymbol{s},\boldsymbol{s}')$. But first, we need consider the BC $c_T(\boldsymbol{s}_D,\boldsymbol{s}')=\overline{\varepsilon_A(\boldsymbol{s}_D)\tilde{T}(\boldsymbol{s}')}$, where the subscript D denotes the Diriclet BC. In this case ($\boldsymbol{s}=\boldsymbol{s}_D$), Eqs. (9)-(10) reduce to

$$\nabla^2 c_T(\boldsymbol{s},\boldsymbol{s}_D)=0, \tag{11}$$

with

$$c_T(\boldsymbol{s},\boldsymbol{s}_D)=\mathrm{cov}(\boldsymbol{s},\boldsymbol{s}_D),\quad c_T(\boldsymbol{s},\boldsymbol{s}_D)\in\Gamma_D. \tag{12}$$

The $\mathrm{cov}(\boldsymbol{s},\boldsymbol{s}_D)$ denotes the covariance between the boundary and the grid points. Such information can be derived from the data or from prior knowledge. For each reference point on the boundary ($\boldsymbol{s}_D$), the corresponding covariances for the entire region are obtained from the Laplace Eq. (11), depending on the BC assumed. Since $c_T(\boldsymbol{s},\boldsymbol{s}_D)=c_T(\boldsymbol{s}_D,\boldsymbol{s})$, the solutions of Eq. (11) can serve as the BC for Eq. (9).

In light of Eqs. (5)-(12) above the *G*-KB should be stochastically formulated in terms of the vectorial function $\boldsymbol{g}$ in Eqs. (1) so that the corresponding expectations are

$$\left.\begin{array}{l}\overline{T}(\boldsymbol{s})\ \text{ for all points } \boldsymbol{s} \text{ in the domain,}\\ c_T(\boldsymbol{s},\boldsymbol{s}')\ \text{ for all pairs of points } \boldsymbol{s},\boldsymbol{s}' \text{ in the domain}\end{array}\right\}\rightarrow\overline{\boldsymbol{g}}, \tag{13}$$

where $\overline{T}(\boldsymbol{s})$ and $c_T(\boldsymbol{s},\boldsymbol{s}')$ are the solutions of the system of Eqs. (5)-(8) and (9)-(12), respectively at all solution grid nodes. These solutions are obtained numerically in section 3.3 below. Explicit expressions of the underlying vectorial function $\boldsymbol{g}$ can be found in the relevant literature (e.g., Christakos, 2000; Kolovos *et al.*, 2002).

3.4. Numerical Solution of Equations (5)-(12)

As we saw above (initially assuming no prior knowledge of the region), the BC of the mean and covariance of the geothermal field can be calculated from the site-specific data using a spatial interpolation method (e.g., kriging or Wiener-Kolmogorov estimators; Christakos, 1992) that generates temperature estimates (and the corresponding variances) at all boundary points of the solution grid.

The temperature mean across space is obtained from Eq. (5) using the Dirichlet BC established in Papantonopoulos and Modis (2005). In this way, the mean is consistent with the heat transfer law and is physically more meaningful than the mean obtained from spatial data analysis with parametric or nonparametric means (Barnett and Turkman, 1997). Similarly, the temperature covariance obtained from spatial data processing only serves as the prior knowledge for Eqs. (11)-(12), i.e., it provides estimates of the relevant Dirichlet BC. Given the covariances obtained from the system (11)-(12), the covariances between any arbitrary grid points $\boldsymbol{s}$ and $\boldsymbol{s}'$ can be derived from (9)-(10). The covariances $c_T(\boldsymbol{s},\boldsymbol{s}')$ describe the spatial dependence of the geothermal field values and are also consistent with the heat transfer law. This is different from the spatial statistics approach that focuses on the data and ignores the underlying physical laws.

Computationally, the identification of the BC for the covariance Eqs. (9)-(10) requires the calculation of a series of BC covariance systems (11)-(12), depending on the reference point $\boldsymbol{s}_D$. As the size of the solution grid increases, the computational effort increases, as well. For illustration, consider a three-dimensional cubic computational grid of size $M \times M \times M$ with $6M^2$ boundary points; the size of each covariance equation system is $M \times M \times M$ and it must be solved $1+6M^2+N_d$ times (for the mean trend and the covariance between the data and the boundary points), where N_d is the number of data. In the Nea Kessani study, a finite difference method (Press *et al.*, 1992) was implemented to solve the mean and covariance temperature (Laplace) equations, whereas the full Multigrid (MG) method (Briggs *et al.*, 2000) was selected as the solver of the Laplace systems (for more details, see Yu *et al.*, 2006).

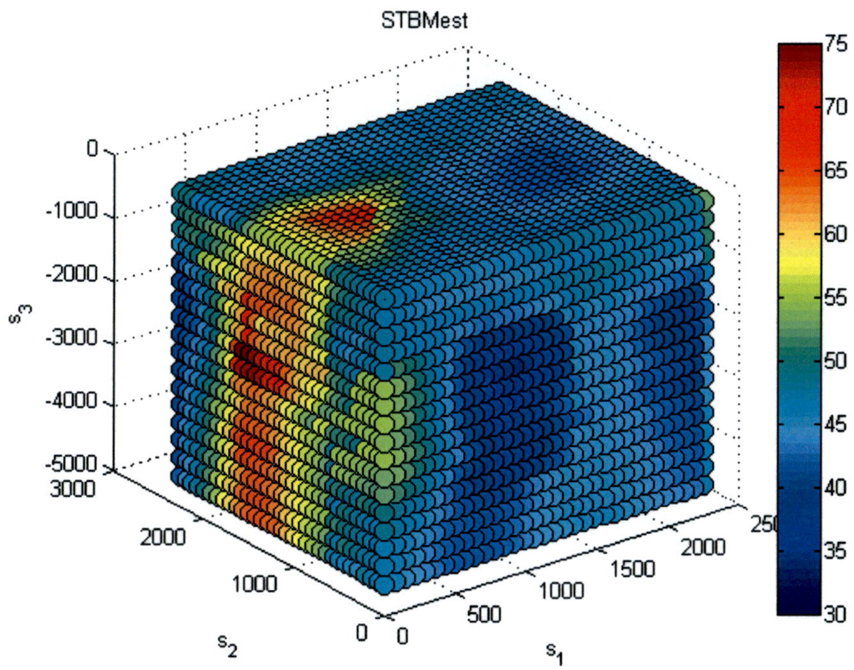

Figure 4. The Dirichlet-type BC for the heat transfer law of Eq. (3); temperature values in $^{\circ}C$.

Eq. (3) was solved for the temperature distribution in the domain of interest using the case-specific BC provided in the earlier study of Papantonopoulos and Modis (2005). The domain was discretized using a $(2^{k+1}+1)\times(2^{k+1}+1)\times(2^{k}+1)$ grid, where k is an integer ≥ 0. For the estimation of the mean temperature across space, the temperature values at 2^{2k+4} grid boundary points were calculated using simple kriging with a nonparametric moving average mean and subsequently served as the Dirichlet BC for the estimation of the temperature at arbitrary grid points. Fig 4 shows the Dirichlet BC values estimated from the data ($k=3$). The spatial distribution of the mean temperature (Fig. 5) was calculated by solving the mean heat transfer equation (5) subject to the BC of Fig. 4. The covariance model is estimated from the residuals--data minus the physical means (5). To construct the BC of Eq. (9) for each grid point, 2^{2k+4} Laplace equations (11) are solved subject to different BC at each boundary grid point. The BC of Eq. (9) are derived from the previously obtained solutions of Eq. (11). Finally, to obtain the solution of Eq. (9) we considered the reference points $\boldsymbol{s}'$ as data points so that the resulting $c_T(\boldsymbol{s},\boldsymbol{s}')$ represent the covariances between all pairs of data points as well as between the solution grid points and the data points.

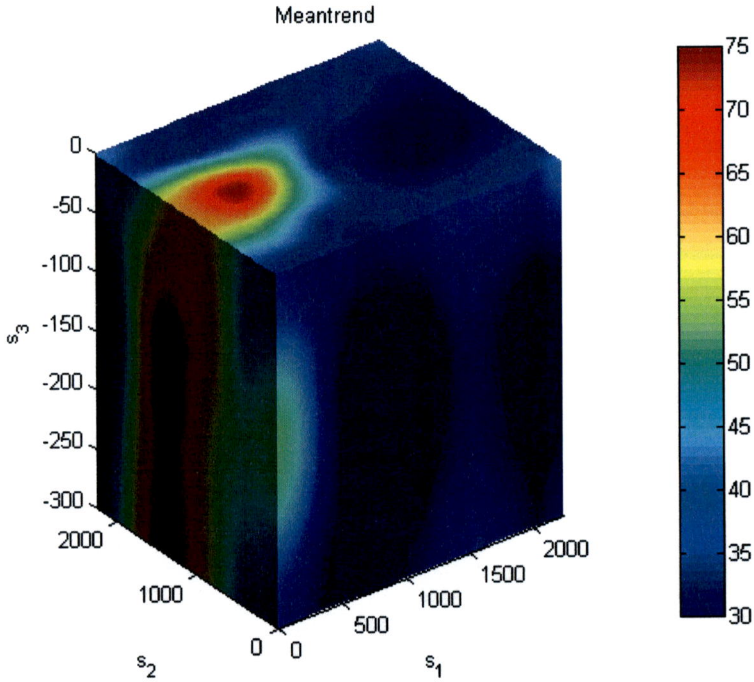

Figure 5. Estimation of the temperature meantrend throughout the computational domain); temperature values in ^{o}C.

4. The Composite (BME) Solution

Given the *G*-KB (13) and the *S*-KB (2), the BME system of Eqs. (1) can be solved for the temperature pdf f_K at each spatial node on the three-dimensional solution grid of the Nea Kessani region. This kind of composite (BME) solution is more complete and informative than the standard ones that provide single temperature values at each solution node (whereas the BME solution provides the complete pdf of possible solutions at each node). The derived f_K may have arbitrary shapes depending on the spatial location, expressing different knowledge and uncertainty levels. Given the f_K, several temperature solutions can be derived, such as the spatial distributions of the corresponding BMEmean temperature and error variance values across space shown in Figs. 6 and 7. I.e., the first map (Fig. 6) is the distribution of temperature solutions of the heat transfer law across space in the BME sense discussed above; the second map (Fig. 7) describes the statistical error associated with the solutions of Fig. 6. The software needed to perform these calculations is the SEKS-GUI software package (see, e.g., discussion in Kolovos *et al.*, 2006).

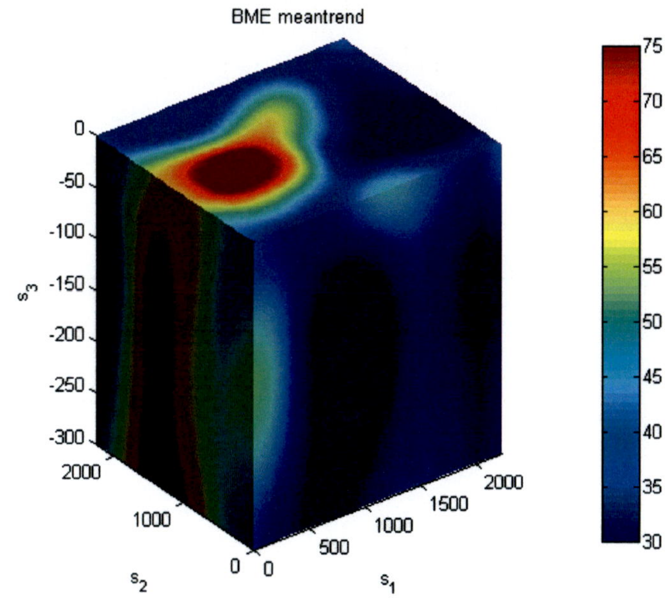

Figure 6. Temperature solution in terms of the BMEmean; temperature values in ^{o}C.

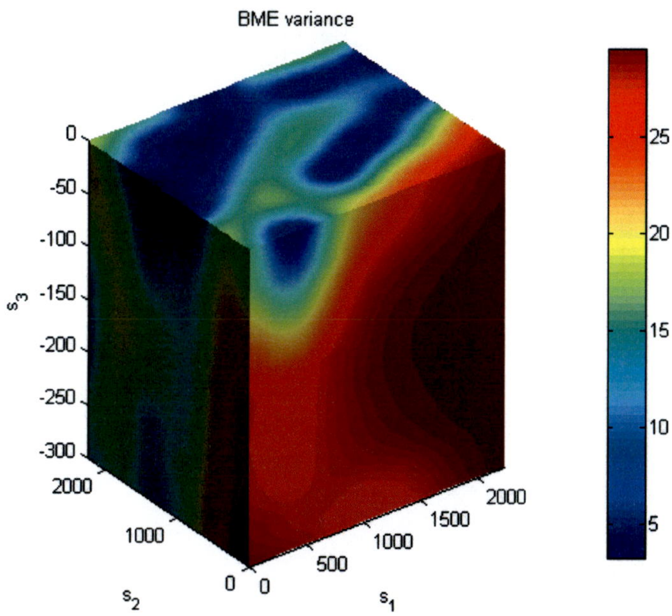

Figure 7. Error variance distribution associated with the solution of Fig. 6; temperature variance in $(^{o}C)^{2}$.

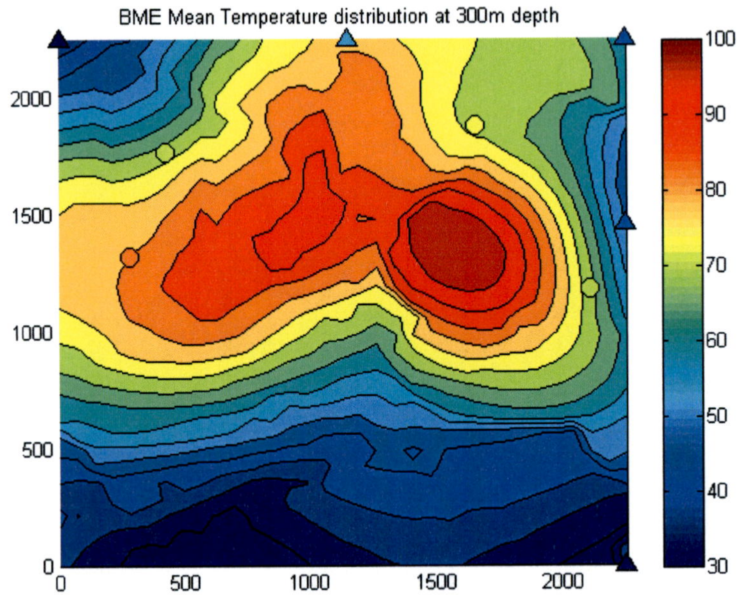

Figure 8. Horizontal temperature distribution (in ^{o}C) at a depth of 300m generated by BME (circles and triangles denote data points).

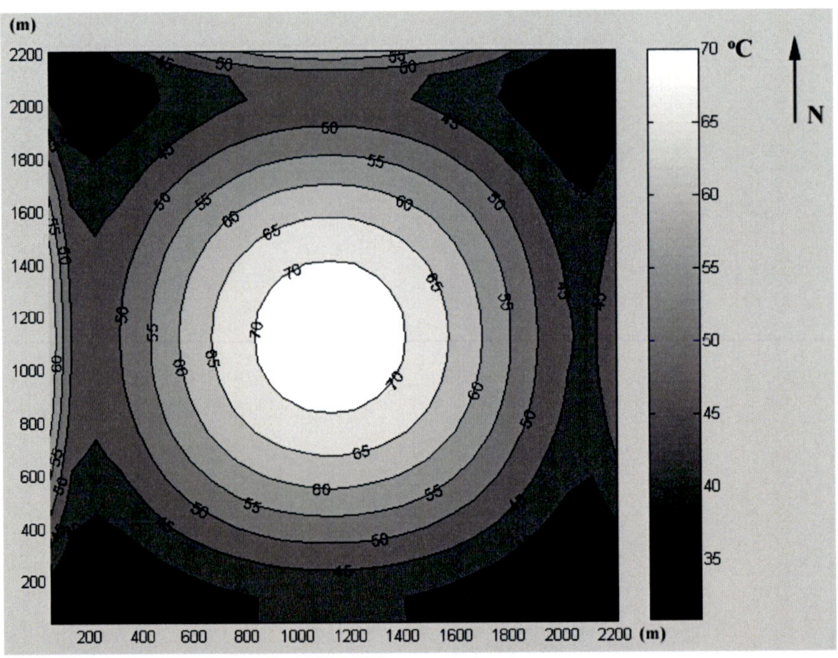

Figure 9. Horizontal temperature distribution (in ^{o}C) at 300 m depth, according to the analytical solution of Eq. (3) with determinitic BC (Papantonopoulos and Modis, 2005).

For illustration, the Figs. 8 and 9 show temperature solutions of the heat transfer equation at a depth of 300m obtained, respectively, by BME and the analytical solution of Eq. (3) obtained using the deterministic Dirchlet BC of the form $\alpha \sin(\pi s_1 / d) \sin(\pi s_2 / g)$, where α

is a constant and *d, g* are the domain dimensions (see Papantonopoulos and Modis, 2005). As should be expected, the use of random BC in the numerical solution of Eq. (3) has improved the obtained results--compare Figs. 8 vs. 9. Moreover, by reviewing all three solutions (i.e. BME, analytical and numerical presented, respectively, in Figs. 8, 9 and 10), it is apparent that an important advantage of the composite BME solution is that in addition to the information contained in the heat transfer law it assimilates the site-specific information contained in the field observations, as well. Clearly, the BME solutions (Fig. 8) can generate temperatures of 110°C, whereas the standard solutions (Figs. 9 and 10) do not exceed the maximum measured temperature value of 80° C. This situation is of some significance, since the 110°C value is in agreement with the maximum expected temperature value according to SiO_2 geothermometry studies as discussed in Grassi *et al.* (1996).

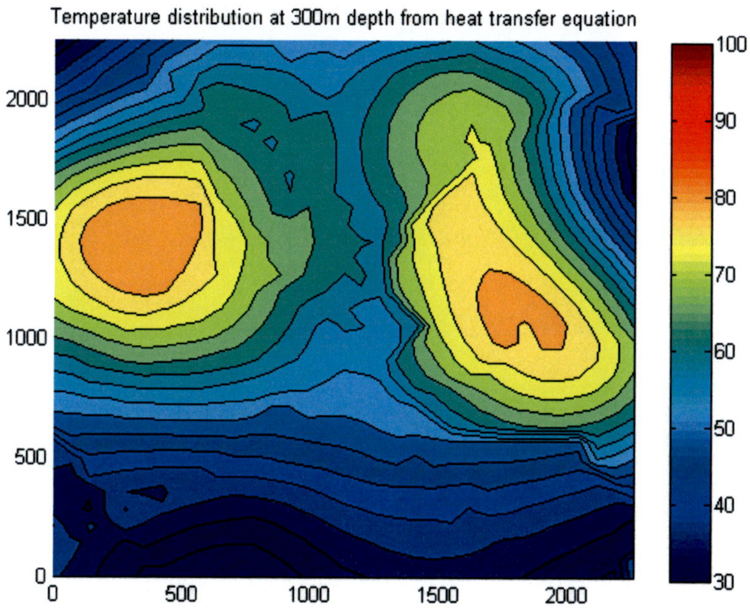

Figure 10. Horizontal temperature distribution (in ^{o}C) at 300m depth obtained from the numerical solution of Eq. (3).

5. Conclusions

The BME approach generates composite temperature maps in a geothermal field domain located in the Nea Kessani region (Greece). The BME approach does not only assure consistency with the physical model, but also accounts for the multi-sourced uncertainty of the model parameters and incorporates site-specific information obtained from in-situ drill holes. In addition, the BME provides the complete temperature pdf at each spatial location. From these pdf, different temperature maps can be derived at all solution grid nodes. The proposed solution is carefully distinguished from the standard (direct) solution of the physical model in a formal mathematical sense. By means of comparative analysis, it is shown that the composite numerical solution is more informative than the direct numerical solution as well as the direct analytical solution obtained using deterministic boundary conditions.

Acknowledgment

The research was supported by grants from the University of Tennessee-Oak Ridge National Lab (OR7865-001.01), and the Fred J. Hansen Institute (Grant No. 54266A P3590).

References

Barnett, V. and Turkman, K. (eds.), 1997. *Statistics for the Environment, Pollution Assessment and Control*. J. Wiley & Sons: New York, NY.

Christakos, G., 1990. "A Bayesian maximum entropy view to the spatial estimation problem", *Mathematical Geology*, **22** (7): 763 - 777.

Christakos, G., 1991. "Some applications of the Bayesian maximum-entropy concept in Geostatistics". In *Maximum Entropy and Bayesian Methods*, W. T. Grandy Jr and L. H. Schick (eds.), 215 - 229, Kluwer Acad.: Norwell, MA.

Christakos, G., 1992. *Random Field Models in Earth Sciences*. Academic Press: San Diego, CA. (New Edition, 2005, Dover Publ. Inc.: Mineola, NY.)

Christakos, G., 2000. *Modern Spatiotemporal Geostatistics*, Oxford Univ. Press New York, NY.

Christakos, G., 2004. "The cognitive basis of physical modelling". Proceed. *Computational Methods in Water Resources (CMWR2004)*, June 13-17, 2004: Chapel Hill, NC, USA.

Christakos, G., 2005. "Recent methodological developments in geophysical assimilation modelling". *Reviews of Geophysics* **43**: 1-10.

Grassi S., N. Kolios, M. Mussi and Saradeas A., 1996. "Groundwater circulation in the Nea Kessani low-temperature Geothermal field (NE Greece)", *Geothermics*, **25**(2): 231-247.

Kolios N., 1993. *Research on the Geothermal Field of Nea Kessani*. PhD Thesis, National University of Athens, School of Mining and Metallurgical Engineering, Athens, Greece. (In Greek.)

Kolovos, A., G. Christakos, M. L. Serre, and C. T. Miller, 2002. "Computational Bayesian maximum entropy solution of a stochastic advection-reaction equation in the light of site-specific information", *Water Resources Research*, **38**(12) 1318-1334.

Kolovos, A., Yu, H-L and G. Christakos, 2006. *SEKS-GUI v.0.6 User Manual*. Interdisciplinary Knowledge Synthesis Group, Dept. Of Geography, San Diego State University, CA.

Papantonopoulos, G. and K. Modis, 2005. "A BME solution of the stochastic three-dimensional Laplace equation representing a geothermal field subject to site-specific information", J. *Stochastic Environmental Research and Risk Assessment*, **20**: 23-32.

Press, W.H., B.P. Flannery, S.A. Teukolsky and Vetterling, W.T., 1992. *Numerical Recipes in FORTRAN: The Art of Scientific Computing*, Cambridge Univ. Press: Cambridge, UK.

Yu, H-L, G. Christakos, K. Modis and G. Papantonopoulos, 2007. "Composite solution of a physical equation representing the three-dimensional geothermal field in Nea Kessiani (Greece)". *Jour. of Geophysical Research.* Accepted for publication.

Zwillinger, D., 1989. *Handbook of Differential Equations.* Academic Press: New York, NY.

In: Geothermal Energy Research Trends
Editor: Herman I. Ueckermann, pp. 19-57
ISBN: 1-60021-683-8

Chapter 2

ADVANCES IN STUDIES OF THERMAL-FLUID GEOCHEMISTRY AND HYDROTHERMAL RESOURCES IN CHINA

***Jianguo Du*[1,2], *Youlian Zhang*[3] *and Heping Li*[2]**
[1] Institute of Earthquake Science, China Earthquake Administration
No.63 Fuxing Road, Beijin 100036, China
[2] Institute of Geochemistry, CAS, Guiyang 550002, China
[3] Seismological Press, China Earthquake Administration
Beijing 100081, China

Abstract

This chapter introduces briefly distribution of hydrothermal resources, potential of hydrothermal energy, geochemistry of geothermal fluids and correlation between geothermal areas and seismic zones in China. More than 3,200 hydrothermal manifestations have been found in China. About 2,240 drilled wells reveal that 275 high temperature sites of hydrothermal energy, which are expected to supply a need of electric generators with total annual output of 5,800 MW. More than 2,900 sites of low and intermediate temperature geothermal systems have been found, which can be utilized for heating, medicine treating, bathing, farming, etc.

Most geothermal waters in China are (Na, Ca)-HCO_3 type, and some are (Ca, Na)-SO_4 and Na-Cl types. Stable isotopic compositions of oxygen and hydrogen indicate the geothermal waters are derived from meteoric water, with small amount of magmatic volatile. Reservoir temperatures calculated with chemical geothermometers range from about 100 °C to 350 °C. Geochemical variations of geothermal fluids with time are found, which are correlated to hydrothermal eruption, earthquakes and exploitation.

Main gaseous components of geothermal systems in China are CO_2, N_2, O_2, and trace amount of H_2S, H_2, CH_4, NH_3, CO, C_2H_6, C_3H_8 as well as noble gases (Rn, He, Ar, Ne, Kr, Xe). The gaseous concentrations of geothermal systems are correlated to the temperatures of geothermal systems and seismic faults. The gases have a multiple origins of crust, mantle and atmosphere.

The chapter emphasizes on both the spatial correlation between the geothermal areas/zones and the seismic zones and the energetic relationship between geothermal-fluid geochemistry and seismic activity. The more amounts of mantle gases the geothermal systems contain, the higher temperature of geothermal systems and the more active the seismic zones. The deep earth fluids provide both matter and energy for geothermal fields and earthquake generation, and carry the messages of geothermal reservoir and earthquake.

1. Introduction

Energy resources are very important to humankinds. The proven commercial reserves of energy stored in all fossil fuels are around 4.673×10^7 PJ (1 PJ = 10^{15} Joules); the present rate of worldwide consumption is nearly 3×10^5 PJ/a, indicating only about 150 years of fossil fuels remain if consumption remains constant[1]. However, the energy consumption in China and other developing countries is rapidly increasing. Otherwise, the constitution of energy consumption in China now, 75% of coal, 17% of petroleum, 2% of natural gas and 6% of others [2], needs eagerly improved. For meeting the increasing need of energy and improving the constitution of energy consumption, it is necessary to find new energy and renewable energy resources. New and renewable energies include geothermal energy, biomaterial energy, wind energy, solar energy, marine energy, hydrogen energy, nuclear energy, gas hydrate etc, except for conventional energy and hydroelectric power [3, 4].

There is a plenty of thermal energy in the earth interior. Four types of geothermal systems have been identified: hydrothermal, hot dry rock, geopressured and magmatic systems. The geothermal system exploited at present is the hydrothermal system. The other three, however, may be industrially exploited in the future after more technological development [3]. Geothermal resources (used here same as hydrothermal resources) include geothermal energy, geothermal fluids and components that can be developed and utilized under the present conditions of technology and economy [5, 6, 7]. The geothermal energy is considered as an important and renewable energy resource. The increased interest in geothermal energy has led to a desire for more sophisticated analyses of available geothermal resources for heat and power production [2, 8, 9, 10]. The geothermal electrical installed capacity in the world was 43,756 GWh/a in 1997 and the direct utilization of geothermal energy was more than 36,910 GWh [11]. In 2000, the geothermal electrical installed capacity in the world increased to 7,974 MW_e, and the electrical energy generated increased to 49,300 GWh annually, representing 0.3 % of the world total electrical energy that was 15,342 TWh, and the thermal capacity in non-electrical usage is 15,144 MW [3]. It is expected that potential of electric power generated by geothermal energy reach 12,000 TWh annually, and direct utilization potential of geothermal energy is about 600×10^9 TWh [4, 11].

This chapter briefly introduces the advances in research on geothermal-fluid geochemistry and geothermal resources in China with emphasis on correlation among geothermal systems, fluid geochemistry, seismic activity and tectonics, as well as structure of lithosphere in geothermal areas.

2. Geothermal Energy in China

Geothermal manifestations, such as hot springs and steam ground, are the phenomena by which humankind began to recognize and utilize geothermal resources. There are large amount of hot springs in China. As early as more 500 years B.C. ago, ancient Chinese recorded utilization of hot springs for medicine treatment. It may be the earliest classification of hot springs that was described in an old book of Song dynasty (420-479 A.D.), in which hot springs were classified into five groups according to smell, color and mineral deposits [6]. The government set up more than 160 thermal-spring resorts in China's mainland in 50th of 20 century, and began to explore insidious geothermal resources at the beginning of 70th of the last century. Geothermal resources have been investigated and exploited in large scale throughout China since then [5, 7, 9, 12, 13]. Exploration and utilization of geothermal energy increased rapidly throughout the world in the fourth quarter of 20 century. As Bertani described, some 1,300 MW of geothermal electrical power generation capacity were installed in 10 countries in 1975, and installed geothermal generation capacity reached 7,974 MW in 2000, indicating addition of 6,700 MW of installed capacity around the world. Up to 1995, the worldwide electrical energy production from geothermal power plants reached about 49,000 GWh/a, but it took a very small part of total electricity consumed worldwide in 1996 (13,700,000 GWh), i.e. geothermal energy accounts for less than 0.4 percent of the world's electricity [14].

Table 1. Worldwide geothermal power generation in early 2005 [14]

Country	Installed capacity (MWe)	Running capacity (MWe)	Annual energy produced (GWh/year)	Number of units	Percent of national capacity	Percent of national energy
Australia	0.2	0.1	0.5	1	Negligible	Negligible
Austria	1.2	1.1	3.2	2	Negligible	Negligible
China	28	19	96	13	30% of Tibet	30% of Tibet
Costa Rica	163	163	1145	5	8.4	15
El Salvador	151	119	967	5	14	22
Ethiopia	7.3	7.3	0	2	1	n/a
France (Guadeloupe)	15	15	102	2	9	9
Germany	0.2	0.2	1.5	1	Negligible	Negligible
Guatemala	33	29	212	8	1.7	3
Iceland	202	202	1483	19	13.7	17.2
Indonesia	797	838	6085	15	2.2	6.7
Italy	791	699	5340	32	1.0	1.9
Japan	535	530	3467	19	0.2	0.3
Kenya	129	129	1088	9	11.2	19.2
Mexico	953	953	6282	36	2.2	3.1
New Zealand	435	403	2774	33	5.5	7.1
Nicaragua	77	38	271	3	11.2	9.8
Papua New Guinea	6	6	17	1	10.9	n/a
Philippines	1930	1838	9253	57	12.7	19.1
Portugal	16	13	90	5	25	n/a
Russia	79	79	85	11	Negligible	Negligible
Thailand	0.3	0.3	1.8	1	Negligible	Negligible
Turkey	20	18	105	1	Negligible	Negligible
USA	2564	1935	17,917	209	0.3	0.5
Total	8933	8035	56,786	490		

n/a: Data not available.

During the first five years of 21st century, the number of countries in which generate electricity from geothermal resources reached 24. Significant geothermal drilling operations were carried out in 19 countries, and 307 new wells were drilled. The total installed capacity worldwide is approximately 8,930 MW_e annually, and electric energy production is nearly 57,000 GWh (early 2005 data, Table 1) [14]. It is clear that utilization of geothermal energy for electricity generation is much smaller. Currently, world energy production is dominated by all types of fossil fuels.

2.1. Geological Setting of Geothermal Resources

China's mainland is a part of Eurasian plate. To the east it borders with the Pacific Ocean plate, and to the southwest it is connected with Indian plate. Taiwan Islands is tectonically located in the collision zone between the Pacific Ocean and Eurasian plates. Most geothermal zones are located in the eastern collision boundary between the India and Eurasia plates and nearby the collision boundary between the Eurasian and Pacific Ocean plates. The different kinds of geothermal fields are formed associated with tectonic and magmatic activities in the different geological periods. China is in the east part of Mediterranean-Himalaya geothermal zone and the west part of the circum Pacific Ocean geothermal zone. There are many Quaternary-Neocene volcanoes in China (Fig.1) [10], which control distribution and property of geothermal resources. However, low-intermediate temperature hydrothermal systems

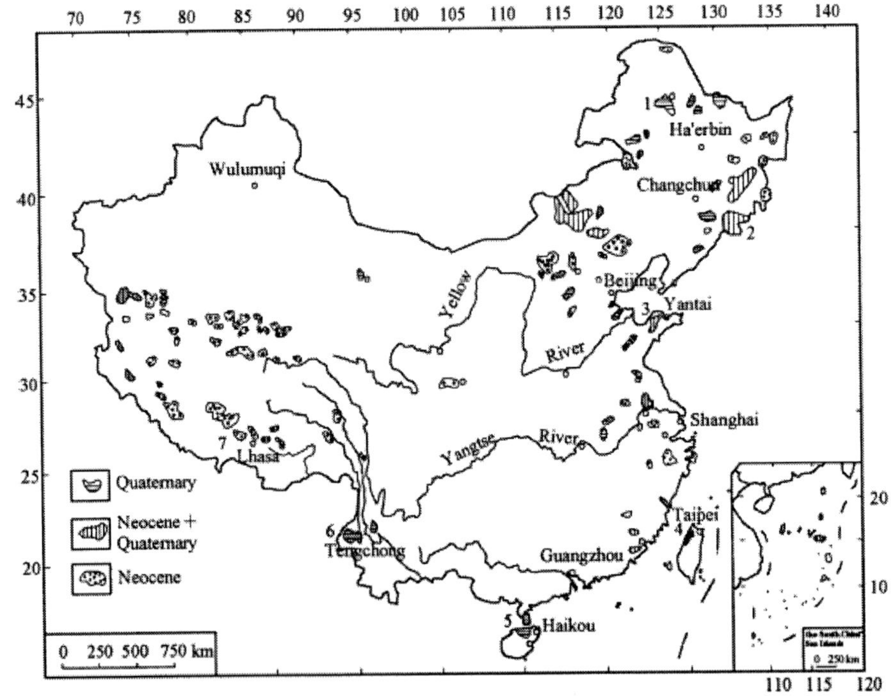

Figure 1. Map of active and recently active volcanoes in China. 1, Wudalianchi; 2, Changbai Mountain; 3, Jiaodong Peninsula; 4, Taiwan Straits; 5, Qiongbei; 6, Tengchong; 7, South Tibet [10].

occurred in some Quaternary volcanic areas in Heilongjiang, Jilin, Inner Mongolia, north China, Leizhou peninsular, north Qiongzhou and Kunlun Mountain, which indicates volcano appears not the unique necessary factor for formation of high-temperature hydrothermal system. About 2200 hydrothermal manifestations with temperature >25°C are found in China before 1994, of which 859 sites are of 25-40°C, 307 sites of >40-80°C, 398 sites of >60-80°C and 136 sites of >80°C. Total heat discharged by the hot springs was estimated to be 101.9×10^{15} J/a [15]. The hot water and steam manifestations are mostly concentrated in Taiwan, southeast coast and Tibet-Yunnan-Sichuan regions, Jiaodong peninsular and Liaodong peninsular. High temperature hydrothermal systems generally occur in orogenes, and low temperature hydrothermal systems in rift-valley or fault-subsidence basins (Fig.2) [9, 12, 16, 17]. Several geothermal zones are recognized in China based on survey and exploration data, such as concentrations of thermal waters, the tectonic characteristics, and geothermal system distribution, lithosphere structure and sporadic warm springs of some districts, as well as the existed geothermal zonings [9, 12, 16]. The geothermal zones are characterized by higher heat flow value (Fig. 3) [18], the active faults (meaning the faults acted or are still acting since 1000 ka before now) [19] and frequent seismic activity (Fig.4). Heat flow values range from 30 to 320 mWm^{-2} with mean value of 63±16 mWm^{-2}, most in a range from 40 to 85 mWm^{-2}. Distribution of heat flow in China is characterized by higher in southeast (70±19 mWm^{-2}), but lower in northwest (43±9 mWm^{-2}); and higher in southwest (81±51 mWm^{-2}), but lower in northeast (61±17 mWm^{-2}) [18, 20]. Chen et al. proposed a systematical classification of hydrothermal systems in China based on geothermal data and the classification principles proposed by Muffler [21, 22].

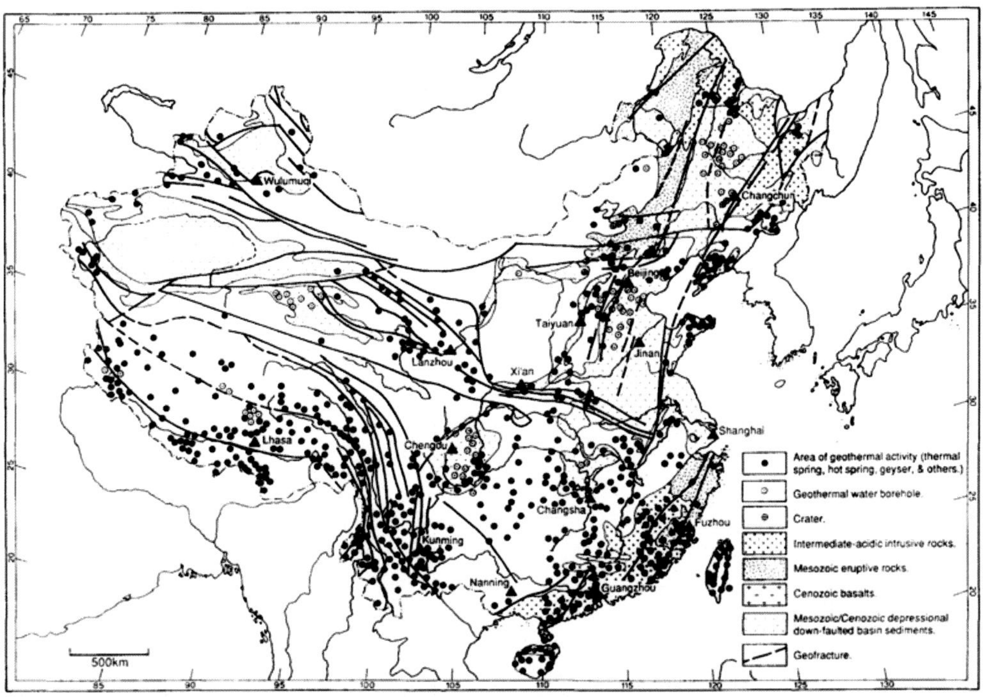

Figure 2. Occurrences of geothermal water in China, showing geothermal manifestation distribution, geothermal areas, simplified geology and major active faults [9].

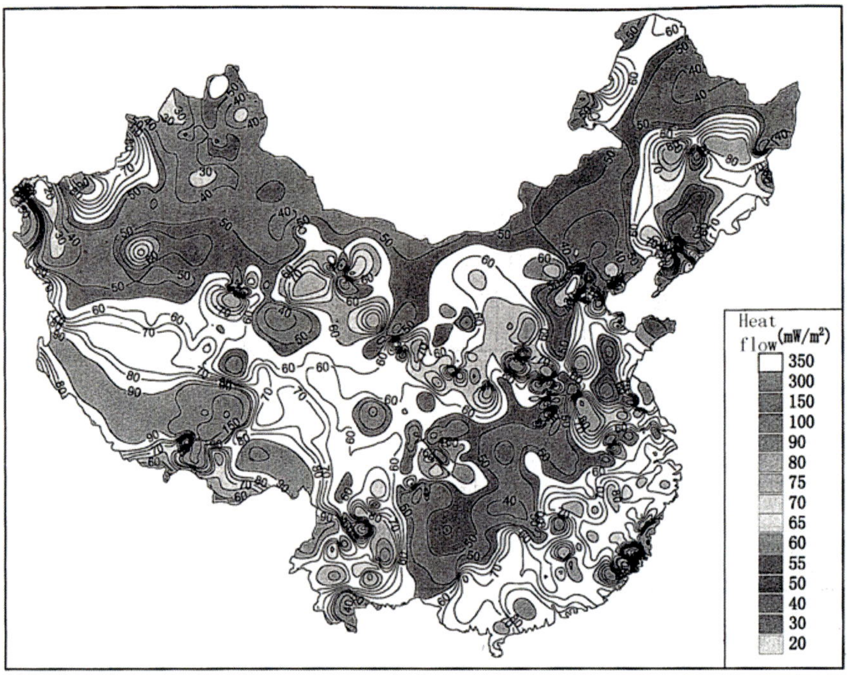

Figure 3. The distribution of heat flow values in China's mainland [18].

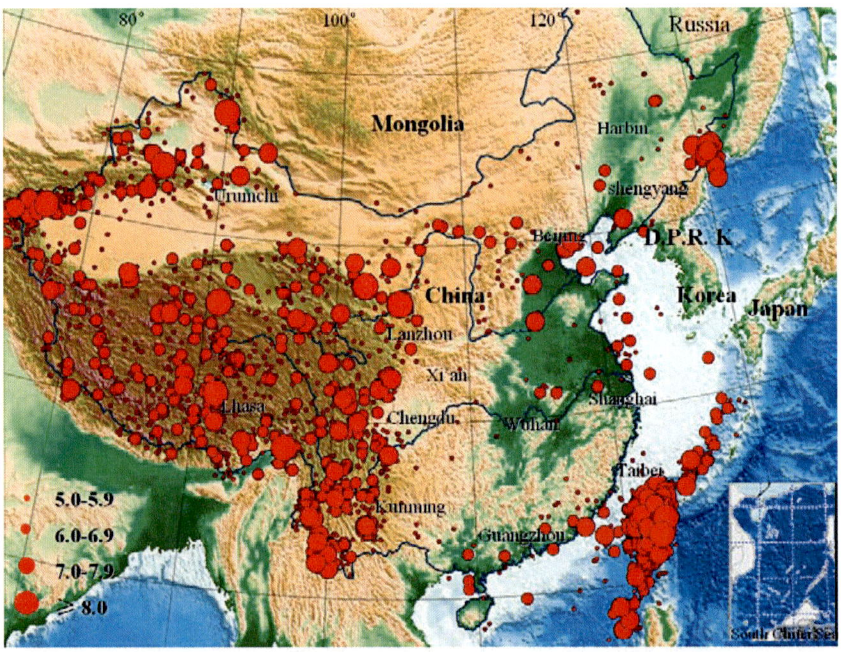

Figure 4. Epicenter distribution of earthquakes with magnitude large or equal to M_S 5.0 in China from 1900 to 2004 (data from the Seismic Database of China Earthquake Administration).

The geothermal areas are characterized by higher geothermal gradient and anomalous heat flow, recent volcanic activity and earthquakes (Fig.1, 2, 3 and 4). High temperature geothermal systems occur most frequently in the active volcanic regions and active tectonic zones. Many such systems are undergoing intensive exploration for energy utilization and a number of systems are already exploited for electricity generation or directly for industrial energy. In the volcanic geothermal systems, temperatures may reach about 350 °C at exploitable depths (<2.5 km) [23]. However, geothermal systems of low and intermediate temperatures are usually found in stable continental environments [9, 12, 24, 25, 26]. The highest heat flow (77 mWm^{-2}) at the continental surface is found in the areas that have experienced magmatic or metamorphic activities more recently than 65 Ma, and heat flow decreases to a constant value of about 46 mWm^{-2} in crust older than 800 Ma; and in the young oceanic crust (<65Ma) heat flow is higher and more variable (70–170 mWm^{-2}) than in older oceanic crust (>65 Ma) that has lower and more constant heat flow (about 50 mWm^{-2}) [3].

2.1.1. The South Tibet-Southwest Yunnan Geothermal Zone

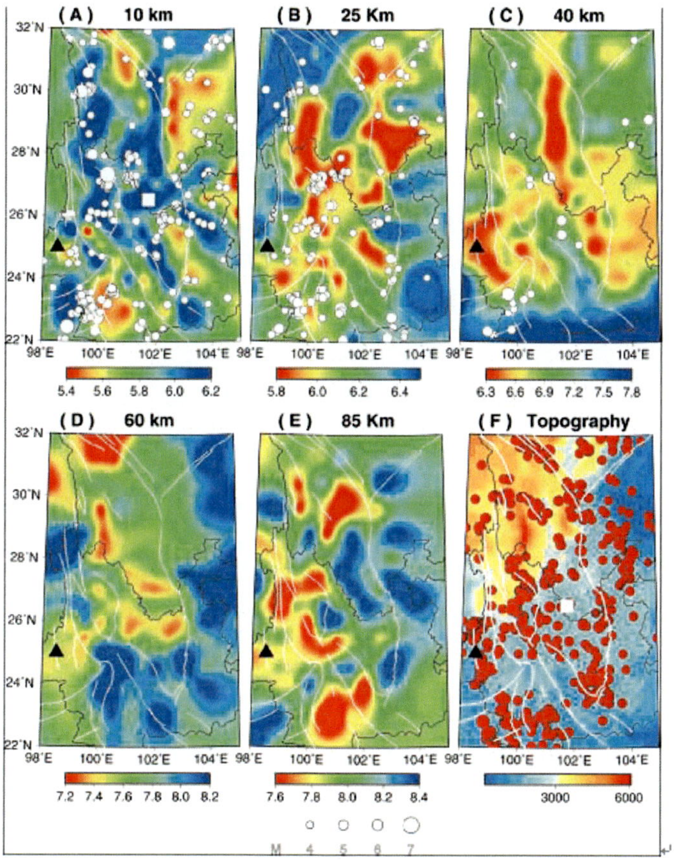

Figure 5. (a–e) P wave velocity at each depth slice. The depth of each layer marked on the top of each map; red and blue colors denote low and high velocities, respectively. (f) surface topography, red and blue colors denote high and low elevations, respectively. White lines are active faults and the black triangle and the white square denote the Tengchong volcanic area and the Panzhihua mining district, respectively. White circles show earthquakes (M > 4.0) that occurred from 1981 to 1998 within a range of 10 km from each depth slice. Red dots show large earthquakes (M > 7.0) that occurred from 1900 to 1998 [29].

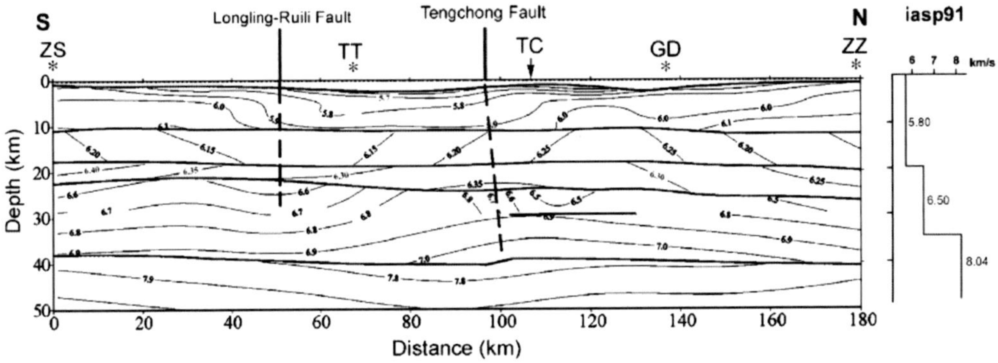

Figure 6. The 2-D crustal structure on N-S main line cross Tengchong volcano-geothermal area (left), and comparison with iasp91 model (right), showing that both the upper crust and upper mantle have negative velocity anomalies, while the lower crust has positive velocity anomaly. There is a local variation of Moho depth beneath Tengchong (TC), which is inferred from the record sections of shots ZZ, GD and ZS. ZS: Zhongshan, 24°05′.2N, 98°35′.6E, TT: Tuantian, 24°42′.0N, 98°39′.9E, GD: Gudong, 25°18′.4N, 98°30′.4E, ZZ: Zizhi, 25°42′.1N, 98°33′.9E [33].

This zone is the largest high-temperature geothermal zone in China and its geothermal activity is the most intense in China's mainland [27]. It extends from the Gangdisi Mountains south of the Nianqingtanggula Mountains eastwards along the Yarlungzangbo River, turns southeasterly at the Nujiang River and enters the Tengchong volcanic region in southwest Yunnan through the Gaoligong Mountain (Fig.2). Over 500 geothermal manifestations have been discovered in this geothermal zone, and a variety of high temperature hydrothermal phenomena, such as hydrothermal explosions, geysers, fumaroles, boiling springs and hot springs, are frequently encountered. There are several active deep faults in the geothermal zone. The largest one is Yarlungzangbo fault that belongs a suture zone between the Indian and Eurasian plates. Magmatic intrusion with a length of about 800 km exists along the northern bank of the Yaluzangbu River. Francheteau et al. reported that heat flow value was about 146 mWm^{-2} near the suture zone in Tibet and dropped to 91 mWm^{-2} in less than 25 km, suggesting the existence of a heat anomaly at depths no greater than 25 km [28]. Magmatic activity and remelting processes in the belt constitute an enormous heat source, and provide excellent channels for hydrothermal activity [9]. In the Tengchong volcanic area, there are Yanshanian granites as well as some Quaternary repeated eruptive volcanics. Neotectonic movement is very active. Due to the continual uplifting of the earth's crust, volcanic activity and earthquakes are very frequent and violent [10]. The high temperature thermal springs in southwest Yunnan are mainly distributed around the Tengchong volcanic zone [9, 30, 31, 32], which may be caused by emission of deep-earth volatiles through deep faults and volcanic framework. The geophysical research has revealed that the low-velocity anomalous zones exist in the upper crust (3-10 km), lower crust (some 40 km) and the top of mantle (about 85 km) in Tengchong area (Fig.5 and 6), in which the Longling–Ruili fault extends from surface to the lower crust and the Tengchong fault likely penetrates the Moho. The 3-D structure of lithosphere in southwestern China shows that both the upper crust and upper mantle have negative velocity anomalies, while the lower crust has positive velocity anomaly. The characteristics of low P-wave and S-wave velocities, low resistivity, high heat-flow value and low Q in the lithosphere suggests presence of magma chamber in the crust derived from the upper mantle, supplying high temperature fluids to the geothermal zone [29, 33]. The estimated total capacity for power generation of the whole zone is about 1,550 MW_e, among which 1,000 MW_e may be expected from south Tibet and 550 MW_e from southwest Yunnan, respectively [21].

2.1.2. The Southeast Coast-Hainan Geothermal Zone

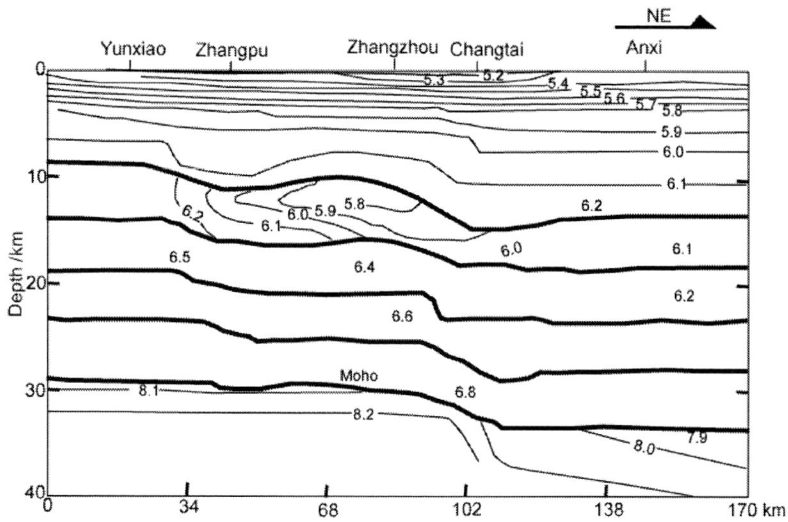

Figure 7. A 2-D structure of lithosphere across the southeast coast geothermal zone. Thick line stands for boundary and thin line is isoline of P-wave velocity [35].

This geothermal zone is located in Yangzi fault block and south China fault-fold zone, mainly in Fujian, Guangdong and Hainan provinces and parts of Jiangxi and Hunan provinces. Granite of different geological ages and Mesozoic volcanic rocks are developed in the geothermal zone, and Quaternary basalt in Hainan Islands (Fig.2). There are several NE-strike deep faults that provide channels for deep-earth fluid migration. More than 460 hot springs are found in this zone and total heat discharged from the springs is $9{,}730\times10^{12}$ J/a. Most of the springs are concentrated in the coastal regions and have higher temperatures. 74% of hot springs in the zone occur in the fault zones in magmatic rocks or contacting zone between intrusive body and host rocks, and 35% of the total number of springs are in Guangdong and Fujing provinces with temperatures over 60°C. Temperatures in the thermal springs and boreholes are slightly lower in the west and higher in the east within this zone. Temperatures of the geothermal water show a northeastwards increasing tendency from the Lianhuashan fracture in Guangdong province to Zhangzhou, Fujian (Fig.2, 3), where several thermal springs have temperatures of more than 90°C; in some thermal areas the water temperatures in boreholes exceed 100°C. The hydrothermal systems in this zone are intermediate temperature, and circulation depths of thermal water are 3.5-4 km [9, 12, 13, 34]. The geophysical data indicate that there is a V_P-lower (5.8 km/s) zone of about 6 km thick beneath depth of 10.2 km in the Zhangzhou geothermal field and its neighborhood, in which P-wave velocity decreases 5-6%. The V_P-lower zone seems to extend around Zhangzhou, indicating a high temperature and fluid-rich zone exists (Fig.7) [35]. Recent geothermic study revealed the geothermal structure of lithosphere in the southeast coast geothermal zone, and shown higher surface heat-flow and higher temperature in lithosphere (Fig.8). There exists a noticeable correlation between the crustal temperature and the regional magnetic anomalies along the Quanzhou-Heshui profile (Fig.8C). Positive and negative anomalies are observed in the low temperature and the higher temperature areas, respectively. This correlation may result from variation of thickness of magnetic crust that is controlled by subsurface temperatures [36].

2.1.3. The Taiwan Geothermal Area

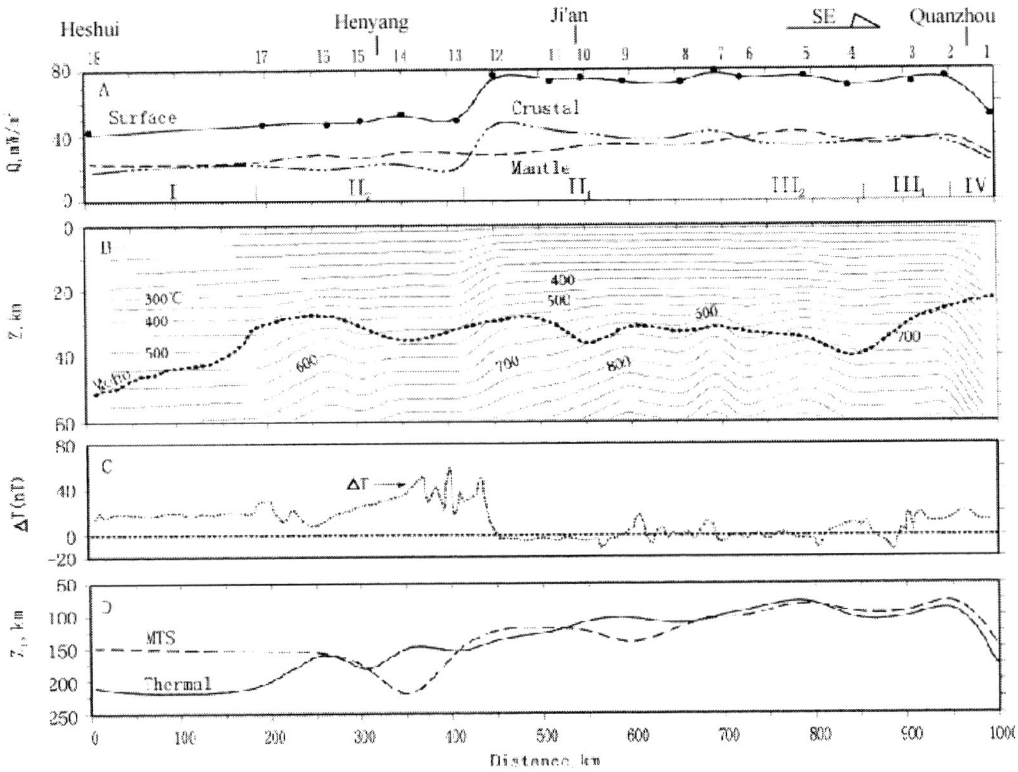

Figure 8. Calculated 2-D temperature distribution (B) along the Quanzhou-Heshui profile together with the observed surface heat flow (solid dots), calculated surface, mantle heat flows and crustal heat contribution (A), and the regional magnetic anomalies (C) as well as calculated ``thermal'' lithospheric thickness compared with the lithospheric thickness derived from magnetotelluric sounding (D). Numbering on top represents the location codes of the selected seismic profile [36].

Taiwan Islands are situated in the junction region of the Ryukyu and Philippine arcs belonging to the Circum-Pacific island arc system. Taiwan orogen is a consequence of ongoing collision since 5 Ma ago between the Luzon volcanic arc along the western margin of the Philippine Sea plate and the passive continental margin of southeastern China. Volcanic activity on Taiwan Island is active in Quaternary Period, same as in Japan. Taiwan geothermal area is both a geothermal region and a seismic one where earthquakes are most frequent and intense [37]. A strong Chi-Chi earthquake (M_w7.5) occurred in 21 September 1999, which caused a surface rupture of about 80 km in length [38]. 103 active hydrothermal sites in this area are found, including six sites where the water temperatures exceed 100 °C. Total heat discharged from the hot springs is $3{,}379\times10^{12}$ J/a [20, 34, 39]. All three main geothermal regions, the Datun, Tuchang-Qingshui and Lushan fields, are high temperature hydrothermal convection type and mostly concentrated in the eastern and western seismic belts. Heat flow values are larger than 80 mWm^{-2} (up to >120 mWm^{-2} in hydrothermal systems) except for those in west coast depression, and mantle contribution is higher [15]. Nearly all the high-enthalpy geothermal zones lie in the regions of post-Tertiary magmatic activity [12]. The volcanic activity and earthquakes may result from migration of deep-earth fluid providing both matter and heat energy to the geothermal systems.

2.1.4. The Sichuan-Yunnan Geothermal Area

The Sichuan-Yunnan (north Chengdu-south Kunming) geothermal area is located in the southern sector of the North-South Tectonic Belt, an active fracture belt (Fig.2). This area extends from Yunnan Province in the south, where it consists of two subzones from Xiaguan to Ninglang in west Yunnan and from Nanpanjiang to Xiao Jiang in east Yunnan, and terminates in Longmen Mountain in northwest Sichuan (Fig.3) [9, 12]. In this geothermal zone, more than 300 hot springs are found, and four large active fault zones and some severe compressive folds form a N-S trend tectonic system. The spatial distribution of hot springs is mainly controlled by the N-S tectonic system. Hundreds of low-temperature springs cluster along the Xiaojiang, Anning and Longmeushan fault-zones, and many high temperature hot springs distribute along Xianshuihe fault. Many thermal springs and earthquakes of magnitude larger than 6.0 occur in three earthquake zones that extend alongside of three fault zones in the western part of Sichuan province [26]. The seismic chromatography shows there are the high temperature and/or fluid-rich zones in lithosphere of the geothermal area (Fig.5) [29], which may control the geothermal and seismic activities.

2.1.5. The Capital Geothermal Area

The capital geothermal area, the region surrounding Beijing in north China, is tectonically located in the north part of north-China block. The north portion of the capital geothermal area is the Yanshan uplift region with E-W strike, the west portion is the Taihangshan uplift regions with NNE strike, the south portion is north China Basin, and to east it connects with Bohai Sea. The north China basin is a large rift-valley basin characterized by alternate uplift and depression. The Yanshan uplift region is relatively stable, in which major tectonic patterns orient in E-W direction. The Taihangshan uplift region has some small intermountain basins. Many active faults of NE–SW strike exist in the North China Basin and the Taihangshan uplift region [40]. Tancheng-Lujiang fault, the largest and longest NNE-strike deep fault in east China, passes through the east part of this area; and Zijingguan fault, as a western boundary of east China rift valley, passes through the west part of this area. The Tancheng-Lujiang fault starts from south of the Yangtzi River, passes northwards through Shandong, Liaoning, Jilin and Heilongjiang provinces, and goes into the Far East of Russia. This deep active fracture is also a seismic belt, with widespread outcrops of Mesozoic and Cenozoic intrusive and extrusive rocks running from south to north. The distribution of the geothermal manifestations in the capital geothermal area is controlled by the fracture zones (Fig.2). Heat flow values (mean value of 62 ± 13 mWm^{-2}) in the capital geothermal area are higher in the uplift region and lower in the depression region. About 50% to 60% of the surface heat flow is derived from mantle, and temperatures at Moho range from 550 °C to 750

°C [15]. The geothermal systems in Jiaodong peninsular, Shandong province, and Liaodong peninsular, Liaoning province, and north China basin are of low and intermediate temperatures [21]. Geophysical research indicate there are some P-wave velocity zones at different depths in the geothermal system regions, which extend approximately in NE-SW direction same as large active faults (Fig.9) [40, 41]. The lower velocity zones may be the sources of heat and fluid of the geothermal systems.

2.1.6. The Qilian-Luliang Arc-Shaped Geothermal Area

This area is mostly located in the northeastern edge of Tibet-Qinghai plateau, includes the Taihang Mountains, Luliang Mountains, Fen-Wei Valley, as well as the Chinling and Qilian Mountains (Fig.2, Xi'an and Lanzhou are in this area). Many NW- and NWW-strike active faults are identified in this area, and many large earthquakes occurred. The famous active fault is Haiyuan fault with about EW strike, along which Haiyuan earthquake with magnitude of 8.5 occurred in 1920, led a lot of death [19]. Hot springs with relatively low temperatures, generally 40-60 °C, are frequently encountered. Some of the springs on the eastern and western edges of the area, for example in the eastern part of the Yingshan Mountains and in Qinghai region, have relatively high temperatures reaching some 90 °C. Thermal water has

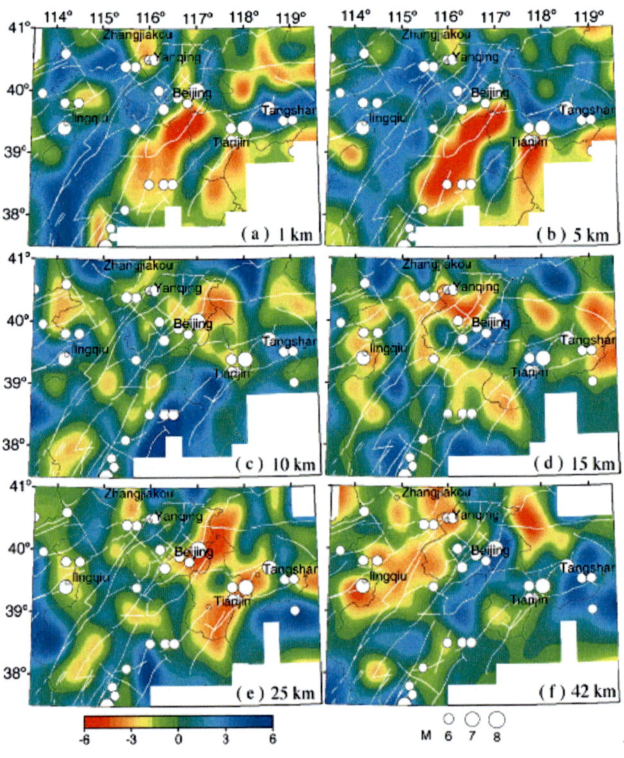

Figure 9. P-wave velocity images at each depth slice (in percent relative to the average velocity). The depth of each layer is shown at the lower right corner of each diagram. Red and blue denote low and high velocities, respectively. White circles show earthquakes (M≥6.0) occurred from BC 780 to 1998. White lines are active faults [40].

been found in some rift valleys and fault subsidence basins, such as the Fen-Wei rift valley, Longdong Basin and the foothill area of Hexi Corridor plain. Ground temperature in Longdong Basin is relatively low, whereas in Xi'an it is about 55°C at a depth of 1000 m. The geothermal systems in the zone belong to the low-temperature thermal water group [12].

2.1.7. West Xinjiang Geothermal Area

This zone, extending NE-SW, is located in the west part of Uigur Autonomous Region along the northwest border of China and characterized by higher heat flow (Fig.3), many geothermal manifestations (Fig.2) and large earthquakes occurred frequently (Fig.4). There is no magmatic volcano in this area, but some mud volcanoes exist in south Tianshan, the north part of this area. A few of work on geothermal exploration was done in the zone because it is a remote area.

In addition, the geothermal energy exploration in recent years revealed that plenty of insidious geothermal energy, both geothermal fluids and hot dry rocks, exists in many other places in China's mainland [2, 7].

2.2. Correlation Between Geothermal and Earthquake Zones

The fluids and heat in the earth are the main factors that control not only geothermic resources, but also volcanic, seismic and tectonic activities. There must be inherent relationships among geothermal resource formation and earthquake generation, distributions of geothermal resource and earthquake. The deep earth fluids as the carrier of the heat upwards migrate, which provides energy for geothermal field, magma generation, volcanic eruption and earthquake generation. Geothermal, fault and seismic activities inversely favor migration of deep-earth fluids. Therefore, fluid geochemical technique can be employed to investigate geothermics and seismicity. Earthquakes frequently occur in the hydrothermal zones that overlap the active tectonic regions. The four global geothermal zones: (1) Circum Pacific Ocean, (2) Mediterranean-Himalaya, (3) Mid-Atlantic ridge, and (4) Red Sea- Aden-East-African Rift Valley zones, are mostly overlap the global earthquake zones [16]. The seven geothermal zones/areas mentioned above are also the seismic areas named as Taiwan and vicinity, Tibet- Sichuan-Yunnna, north China, Guangdong-Fujian coast, and northwest China seismic areas in which 23 earthquake zones are certificated. Comparing the map of hot spring distribution (Fig.2), map of heat flow (Fig.3) and map of geothermal zone [9] with the active fault map [19] and seismic map (Fig.4) in China, it is clear that there are large active fault zones in the geothermal zones/areas where seismisity and volcanism are very active. Geothermal manifestations occur in earthquake areas in China, and large earthquakes frequently occurred in the geothermal zones [26, 42]. Additionally, hot springs are well known in active tectonic zones associated with plate convergence, as exemplified by the European Alps, Rocky Mountains, Himalaya, the New Zealand and Southern Alps [43]. Geothermal areas and associated seismic activity are well known from the central and northern Aegean, and seismic events resulted in massive expulsions from crust [44]. Hydrothermal explosion and earthquakes frequently occurred in the Rehai geothermal field, Yunnan [45]. The more mantle heat in the active fault zones, the higher temperatures of geothermal areas, and the more active the earthquake activities in southwest China [26].

Helium isotopic ratios of hot springs in earthquake zones in southwest China indicate that the gasses derived from mantle carry heat to the crust [46]. The studies of the temporal characteristics in seismic and volcanic activity and relationship with seismogenic stress in East Asia, mainly based on the qualitative and statistical analysis of historical and instrumental earthquake data, indicate that a long-term synchronous variation in seismicity since 1400 A.D. exists in the intraplate region from northeastern China to the Inner Zone of Southwest Japan through the Korean Peninsula, and short-term synchronous variation in seismicity since 1900 is also found in the region. Historical volcanic activity in and around the Korean Peninsula is closely correlated with active seismicity [47]. Tong et al. reported that earthquakes with $M_L \geq 3.0$ occurred in the places of Hunan province where hot springs distribute nearby active faults after 1960, but few earthquakes occurred in no hot-spring regions [48]. In the capital geothermal area, earthquakes are concentrated in four seismic zones. The Zhangjiakou–Penglai seismic zone, oriented in NW–SE direction, is the most active one that contains a majority of large earthquakes and some hydrothermal systems [15, 49]. Others (Tangshan–Xintai, Sanhe–Linshou, and Huailai–Weixian seismic zones) are all oriented in NE–SW direction and are generally parallel to large fault zones. There are some P-wave velocity zones in the crust at the same locations of hydrothermal systems (Fig.9) [37, 40]. The focus of three large earthquakes, 15th Novvember 1976 Ninghe M6.9, 28th July1976 Tangshan M7.8 and Luanhe M7.1 earthquakes, occurred nearby lower velocity zones (Fig.10) [41]. The 4th February 1975 Haicheng M7.3 earthquake was also generated near the top of a lower-velocity and high-resistance zone. Moreover, Jiang et al. obtained the similar distribution images of earthquakes, ground temperature and precipitation in China by statistically analyzing the independent data, showing that the four zones of high seismic frequency circularity areas approximately distributed in a WE trend and well overlapped with the four high-value axes of mean square deviation of ground temperature at depth 3.2 meters, and that the frequency anomaly of precipitation (R> 50%) is distributed in the same zones. This may be attributed to coupling between atmosphere and lithosphere [50].

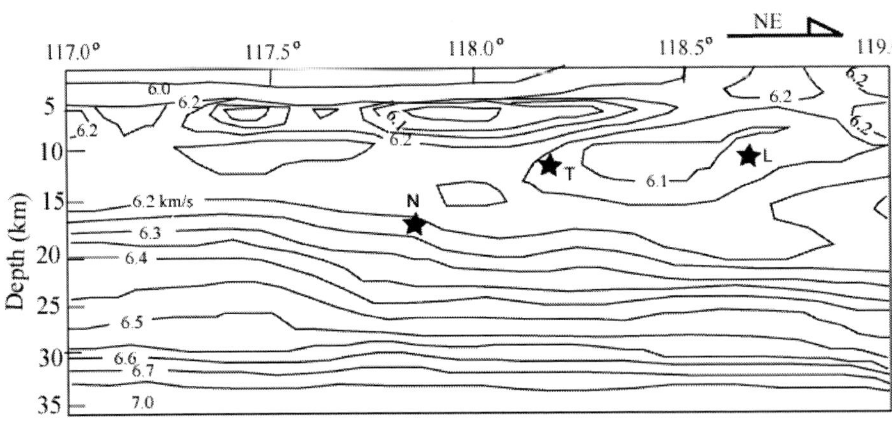

Figure 10. Diagram of Crustal structure on NE profile across Tangshan, absolute velocity isolines display a lower velocity zone nearby which three earthquakes occurred, stars marked N, T and L indicate sources of Ninghe, Tangshan and Luanhe earthquakes, respectively [41].

2.3. Geothermal Energy

2.3.1. The Features of Geothermal Energy in China

The geological survey proved a plenty of geothermal energy potential, showing China is rich in geothermal resources. The proved reserves of geothermal energy in China, however, are not so much because the limited geothermal exploration has been conducted. An annual report about geothermal resources and utilization in China, written by Department of Hydrological and geological Environment, Geological Survey of China according to the update data in the beginning of 2006, described the geothermal energy potential as well as the utilization of geothermal energy in China [7]. The report comprehensively analyzed geothermal resources, presented the assess amount of geothermal energy, put forwards suggestions for exploitation, utilization of geothermal energy as well as protection for the resources and environment, in order to provide the basic data for the state and local governments.

The characteristics of geothermal resources in China are summarized as following [2, 6, 7, 27, 39, 51]:

A The geothermal resource distribution evidently shows a tendency of variation. In west China, the highest heat flow values (91-364 mWm^{-2}) were measured in southwest China, especially in the Yalungzangbu suture zone; and decreases gradually northwards to the values of 33-44 mWm^{-2} in Juger basin in Uiger Autonomous Region. In east China, heat flow values are higher in Taiwan block (80-120 mWm^{-2}), and in southeast coast area (60-100 mWm^{-2}); and decreases northwards to 57-69 mWm^{-2} in Jianghan basin in Hubei province (Fig.3) [7, 13, 18, 25].

B The geothermal resources can be classified into five genetic types: (1) recent volcanic type, with high temperature ($T \geq 150°C$); (2) magmatic type, with high temperature; (3) fault type, typically intermediate temperature ($150°C > T \geq 90°C$); (4) depression basin type, with low ($25°C < T < 90°C$) and intermediate temperature; and (5) fault-subsidence basin type with low and intermediate temperature.

C The conduction type of geothermal energy is mainly intermediate temperature, and is found in the sediment basins. The thermal reservoirs are predominantly Tertiary sandstone and sandy conglomerate with thickness from several hundreds to thousands of meters [5, 7].

D The convection type of geothermal energy is found in the mountain areas. High temperature geothermal energy is mostly found in Tibet, Yunan and Taiwan Islands. The geothermal energy of low and intermediate temperature is found in southeast coat region, north China and northeast China.

2.3.2. Assessment of Geothermal Energy

According to the state standard, "The Exploration Regulation for Geothermal Resources" (GB11615-89), both geothermal energy in thermal reservoir and geothermal fluids are assessed. Geothermal energy in reservoir (J), viable hot fluid reserve (m^3/d or M^3/a) and usable heat (J) were calculated when temperature of hot spring is equal and lager than 25°C in

Table 2. The assessment of usable reserve of geothermal fluids [7]

Location	Area $\times 10^4 km^2$	Mountain area		Sediment basin		Total	
		Hot water ($\times 10^{12} m^3/a$)	Heat ($\times 10^{12} kJ$)	Hot water ($\times 10^{12} m^3/a$)	Heat ($\times 10^{12} kJ$)	Hot water ($\times 10^{12} m^3/a$)	Heat ($\times 10^{12} kJ$)
Beijing	1.68	31.2	0.026	8728.3	9.757	8759.5	9.783
Tianjing	1.1	34.5	0.36	6531.0	13.56	6385.5	23.596
Hebei	19	887.6	1.003	42990	73.796	43877.6	74.799
Shanxi	15	952.5	0.785	10350	10.833	11302.5	11.591
Inner Mongolia	110	2125.8	2.973	117400	113.452	119525.8	116.025
Liaoning	15	1875	2.473	7500	7.85	9375	10.323
Jilin	18	307.8	0.366	23220	24.304	23527.8	24.67
Heilongjiang	46	909.9	0.762	47010	49.205	47919.9	49.967
Jiangsu	10	907.5	1.227	11850	19.349	12757.5	20.576
Shanghai	0.85			1740	2.55	1740	2.55
Zhejiang	10	750	0.55			750	0.55
Anhui	13	1221	1.125	14580	20.144	15801	21.269
Fujian	12	10800	18.087			10800	18.087
Jiangxi	16	7200	5.668			7200	5.667
Shandong	15	1278	1.68	19440	42.73	20718	44.41
Henan	16	1830	1.908	29700	39.791	31530	41.699
Hubei	21	7743.8	5.025	9750	14.287	17493.8	19.312

Table 2. Continued

Location	Area $\times 10^4 km^2$	Mountain area		Sediment basin		Total	
		Hot water ($\times 10^{12} m^3/a$)	Heat ($\times 10^{12} kJ$)	Hot water ($\times 10^{12} m^3/a$)	Heat ($\times 10^{12} kJ$)	Hot water ($\times 10^{12} m^3/a$)	Heat ($\times 10^{12} kJ$)
Hunan	21	11025	5.862			11025	5.862
Guangdong	18	12705	23.937	3180	7.19	15885	31.127
Guangxi	23	2062.5	2.176	250	0.341	2312.5	2.49
Hainan	3.4	2610	3.923	3000	3.14	5610	7.063
Sichuan	48	2936.7	6.037	46110	77.221	49146.7	83.258
Chongqing	8.23	1680	1.02	7890	9.91	9570	10.93
Yunnan	38	34139.7	38.021	600	1.005	43739.7	39.026
Tibet	120	72000	191.722			72000	191.722
Shaanxi	19	1365	1.52	29700	36.683	31065	38.203
Gansu	39	1573.8	1.252	6385	6.683	1958.8	7.935
Qinghai	72	1753.8	2.078	6770	6.519	8523.8	8.597
Ningxia	6.6	123	2.86	1250	1.308	1373	1.411
Xinjiang	160	2511	2.86	38150	36.737	40661	39.597
Taiwan	3.6	3780	9.401			3780	9.401
Total		190650.1	334.763	493894.3	627.918	684544.4	972.681

the mountain area and geothermal gradient is equal and lager than 3°C /100m at depth shallower than 2,000 m in the prospecting geothermal area. According to the annual report, the geothermal reserve is estimated as 73.61×10^{20} J in the sediment basins. The usable reserve of geothermal fluids is estimated as 6.8×10^{9} m^3/a, containing heat is equal to 963×10^{15} J. The detail assessment is listed in Table 2 [7], showing the distributions of geothermal energy in the different parts of China. Up to date, 275 geothermal fields are found and 2,239 exploration wells for geothermal energy are drilled, which are mainly distributed in north China, Shaanxi, Shanxi, Yunnan and southeastern coast area. Among the high temperature geothermal systems in China, 255 systems can be utilized to produce electricity energy with a total potential energy of 5,800 MW, and 10 geothermal systems have been exploited with a production capacity of 300 MW [2].

2.4. Geothermal Energy Utilization

The first success producing electricity from geothermal water was in 1904. Up to 2005, electricity energy are produced from geothermal energy in 24 countries with a total output of 56.95TWh/a, and 75.94 TWh geothermal energy is directly used annually in 72 countries. New and renewable energy, such as geothermal energy, solar energy, wind energy, biomaterial energy and tide energy, take a very small part of global energy construction. The four new and renewable energies were globally estimated to have large potential. The statistics indicates that direct applications of geothermal energy throughout the world increased in last ten years, and that the consumption of geothermal energy for different usages varied relatively (Table 3) [7].

Table 3. The relative consumptions of geothermal energy for direct applications in the world [7]

Time	1995	2000	2005
Heat supply	33%	37%	20.1%
Bathing & amusement	15%	22%	28.8
Fish culture	13%	7%	4.2%
Heat pump	12%	14%	33.2%
Green house	12%	12%	7.5%
Industry	10%	6%	4.2%
Agriculture	1%	1%	0.8%
Melting air conditioner	1%	1%	0.7%
Others	3%		0.4%

Geothermal energy has been used for a long time in China, but up to date utilization of geothermal energy is still small comparing with the total energy consumption, especially for producing electricity from geothermal energy. It is recently estimated that more than 700 sites of hot springs are exploited, and only four sites are used for producing electric energy. The statistical data from Geothermal Division of China Energy Association show, up to 2005, geothermal energy consumption for direct applications is 12,604.6 GWh/a. The geothermal energy consumption for direct applications in China is the highest in the world according to the report on the International Geothermal Conference of 2005 (Table 4) [7]. The direct applications of low and intermediate geothermal energy in China are consisted of 18.0% for heat supply, 65.2% for medicine treatment and swim, 9.1% for agriculture and fish culture and 7.7% for others. The installation capacity of electricity energy produced from high temperature geothermal energy was 32.08 MW in China at the end of 1999, the detail information is shown in Table 5 [7].

Table 4. The first ten countries for direct utilizing geothermal energy in the world [7].

No.	Country	Produce capacity
1	China	12604.6
2	Sweden	10000.8
3	U.S.A.	8678.2
4	Iceland	6615.3
5	Turkey	5451.3
5	Austria	2229.9
7	Hungary	2205.7
8	Italy	2098.5
9	New Zealand	1968.5
10	Brazil	1839.7

Table 5. The installation capacity of electric energy produced from geothermal energy in China [7]

Location	Site	No. of generator	capacity (MW)	Output (MW)
Tibet	Yangbajing	9	25.18	24.18
	Naqu	1	1	Stopped in 2000
	Langjiu	2	2	1
Guangdong	Fengshun	1	0.3	0.3
Hunan	Huitang	1	0.3	0.3
Taiwan	Qingshui	1	3	Stopped in 1995
	Tuchang	1	0.3	Stopped
Total			32.08	25.78

The exploitation and utilization of geothermal energy, however, also resulted in bad effects on environment and construct engineering, to which attention must be paid before and during exploitation of geothermal water. Ground depression occurred in the region where a large amount of geothermal water is exploited. For example, ground depression was caused by pumping geothermal water in Yangbajing region, Tibet, which has been controlled by taking the suitable measurement. A lot of tail hot water produced by direct utilization of geothermal water is harmful if it is not treated before discharge. Direct discharge of tail hot water to land results in temperature increase of land, salting of soil, chemical pollution of environment, vegetation damage etc. Therefore, it must be emphasized that the proper measurements should be taken for utilization of geothermal water in order to protect both hydrothermal resource and environment.

3. Geochemistry of Geothermal Water in China

Geochemcal techniques are widely used in the geothermal research in order to investigate chemical characteristics and origin of geothermal fluids, estimate reservoir temperature, and assess geothermal resources. The major element geochemistry of geothermal fluids is predominately controlled by a set of temperature-dependent mineral-fluid equilibria although chloride and rare gas contents appear to be independent variables reflecting the sources of these components [23], so that many chemical geothermometers are established based on the equilibria [52, 53, 54, 55, 56, 57, 58]. A lot of geochemical research work on geothermal investigation was carried out in China in recent years. Consequently, large amount of geochemical data are obtained, and some new geothermal areas are found.

3.1. Chemical Compositions of Geothermal Waters

Chemical types and mineralization of geothermal waters reflect the geochemical processes of water-rock interaction and reflect chemical component of rocks in which waters circulate. Most of geothermal waters in China are Na-HCO_3 and Ca-HCO_3 types; some are Ca-Mg-SO_4, Na-Cl-HCO_3, Na-Cl-SO_4 and Na-K-Cl types. The geothermal waters occur in sedimentary and metamorphic country rocks are usually Na-HCO_3, and ones occur in igneous/volcanic country rocks are Na-SO_4, Na-Cl-SO_4 and Na-Cl. For example, most geothermal waters found in metamorphic and sedimentary rocks in Taiwan are of Na-HCO_3 with intermediate or larger pH values; but small amount of geothermal waters occurred in volcanic rocks are of Na-SO_4, Ca-SO_4 and Na-Cl types with acidic property [59]. Geothermal waters from Mesozoic fractured granite reservoir with a depth about 300 m in the Nantian geothermal field in Hainan Island are typically Na-Ca-Cl type water, which are characterized by total hardness from 342 to 746 mg/L, pH values of 7.21-8.35 and rich in F (2.07-7.7mg/L), H_2SiO_3 (68.5-117.5 mg/L) and strontium. Geothermal waters of hot spring and wells in fractured granite, monzonite, diorite and metamorphic rocks in Xinzhou geothermal field, Yangdong county, Guangdong province, are Na-Cl type water [60]. More than fourteen hot springs occur in granite, granodiorite and Protozoic metamorphic rocks along some active faults in Shandong Peninsular, Shandong province. Cations in the hot spring waters, with temperature of 50-92 °C, are mainly Na and Ca, and anions are different amount of HCO_3, SO_4 and Cl.

The hot spring waters are Na-Ca-HCO_3-SO_4-Cl, Na-Ca-HCO_3-Cl-SO_4, and Na-Ca-C1-SO_4-HCC_3 types, indicating fractures in the fault zones provide passway for deep fluid migration upwards [61]. Geothermal groundwater found in Cambrian limestone in the Longmen geothermal field, Luoyang county, Henan province, is characterized by mineralization of 398-410 mg/L, Ca-Mg-HCO_3 type, low temperature (24-42°C), and decreasing Cl and SO_4 concentrations with increasing distance from fault [62]. Hot spring waters in Jiangxi province are (Na, K)-HCO_3 type waters with TDS higher than 1000 mg/L [63]. Some 100 hot springs exist around Tianchi Volcano at the border between China and D.P.R. Korea with high temperature about 82□. The thermal spring waters with pH values of 6.9 to 7.1 are Na-HCO_3 type [64, 65]. Thermal water in the geothermal fields in Beijing are predominantly Na-SO_4-HCO_3 or Na-Ca- HCO_3 type waters with mineralization of about 500-700 mg/L and moderate or slight alkaline pH values (7.1-7.9). The well No.Jingre-42 in Beijing urban revealed that thermal water has TDS of 486 mg/L and is Ca-Na-SO_4-HCO_3 type water with isotopic age (^{14}C) of 19,400±330 a [66]. Additionally, geochemical analysis of geothermal waters in Hunan province indicates that geothermal waters are mainly bicarbonate (Na-HCO_3 and Ca-HCO_3). The geothermal water of Ca-HCO_3 type, with mineralization <0.5 g/L, hardness <16.8°, pH>8.2 and dissolved CO_2 up to 6.6-35.86 mg/L, distributes along faults in limestone area [67]. Whereas, the geothermal water of Na-HCO_3 type is characterized by mineralization <0.4 g/L, hardness <17.9°, SiO_2 up to 110 mg/L and higher radon content, and distributes in granite area and the contact zone. Some Ca-SO_4 type geothermal waters with strong H_2S smell and pH of 7.8 are found in Ordovician and Carboniferous strata in the west part of Hunan [68].

Chemical compositions of geothermal waters are affected by water temperatures. For example, the cool waters found in the Yangbajing geothermal field and vicinity are Ca-HCO_3 type; the hot spring waters with lower temperature are Na-Cl-HCO_3 type, which are mixture between cool water and hot water derived from shallow reservoir; Ca-Mg-SO_4 type waters occur in fumaroles; and Na-Cl type waters are encountered in the high temperature springs, boiling springs and deep reservoirs [69]. Ion concentrations of geothermal waters with intermediate temperatures (31-43°C) in the Wuyi geothermal field, central Zhejiang province, are two times of those of local surface water and higher concentrations of F and SO_4 are attributed to fluoritezation and pyritization. The thermal waters are Na(Ca)-SO_4-HCO_3 type [70]. The concentrations of Na, K, HCO_3, SiO_2, Cl, and SO_4 of geothermal waters from the depths of 1250-3000 m in Xi'an geothermal field, Shaanxi province, show an approximately positive linear correlation with wellhead temperature, whereas Ca and Mg vary independently due to ion exchange [71]. Lower temperature waters in the Wudalianchi volcanic area, Heilongjiang province, is bicarbonate type, and SO_4 and Cl concentrations increase in deep artesian water [72]. Dozens of hot spring waters in the Rehai geothermal field, Yunnan province, have been analyzed and can be classified into three chemical groups: (1) Na-Cl-HCO_3 waters discharged from the hottest springs, (2) Na- HCO_3-Cl waters mainly found in the hot springs adjacent to springs of the first type, and (3) Na-HCO_3 waters mainly found in lower temperature springs [32, 73].

Chemical components of geothermal waters usually very with aquifer depth which may reflect water maturity. For example, the hydrogeochemical data for waters from geothermal production wells with depths of 300-3000 m in the Xi'an geothermal field indicate TDS from 420 to 5,033 mg/L and slightly alkaline pH values (7.4 to 8.4). The major ions are in the following order of abundance: Na>Ca>Mg and Cl, SO_4>HCO_3. The geothermal waters change from HCO_3 dominant to Cl dominant with increasing depth of reservoirs, a trend towards maturity. The geothermal waters with the highest temperature and salinity occur at the greatest depth [71]. TDS and Cl in thermal water with moderate temperature (58-86 °C) from 500-1200 m underground increase with increasing depth, and the thermal water is Na-HCO_3 water in shallow reservoir and becomes Na-HCO_3-Cl type in deep one in the Xiong county geothermal field in Hebei province [8]. Hot spring waters in the Rehai geothermal field are Na-Cl-HCO_3 type waters with higher concentrations of SiO_2, Cl and Li, but lower Ca and Mg that are usually found in the high temperature geothermal system [54, 57], indicating the deep source water of the hot springs in the Rehai geothermal field may be Na-Cl water [27, 32]. The same phenomena in the Fuzhou geothermal field in Fujian province are also found [13]. Geothermal waters with temperature of 25-80°C from Neogene (N) sandstone and silt-sandstone aquifers at depths from 450 m to 2,500 m in Tianjing are Na-HCO_3 type water in the northeast district, and become Na-Cl-HCO_3 type water in the southwest district. The chemical types of thermal waters change from Na-HCO_3 at shallow reservoir in Minghuazheng Group (N_m) to Na-Cl-HCO_3 type in the deep reservoir in Guantao Group (N_g), indicating mineralization and maturity of thermal water enhanced from shallow to deep [74].

The ion concentrations in geothermal waters vary with time and variations of temperature and gas components. From 1984 to 2001, bicarbonate and SiO_2 in geothermal water form the well No.Jingre-42 increased by 0.7 and 0.2 mg/L, respectively, and water temperature varied slightly, demonstrating that deep thermal water supply increased during exploitation of the thermal water [66]. The geothermal and geochemical data indicate that reservoir temperatures and ion and gas components of hot springs in the Rehai geothermal field varied from 1973 to 2000, which resulted from time variations of contributions of mantle and shallow crust fluids companying with hydrothermal eruptions [32, 45]. Majumdar et al. reported that both non-thermal and thermal waters from the Bakreswar and Tantloi geothermal areas, located in the Birbhum district of West Bengal and Santhal Parganas district of Bihar, respectively, underwent seasonal isotopic variations, being enriched in ^{18}O and D in the winter relative to the rainy season, and the $\delta^{18}O$ and δD values of non-thermal and thermal waters indicate a meteoric origin for the thermal springs [75].

The geothermometers are generally used in geothermal exploration to estimate temperatures of thermal reservoirs. The estimated temperatures of thermal reservoirs of most geothermal fields in China, at depths from several hundreds to three thousands of meters, are about 100 °C to 300 °C [32, 60, 64, 76, 77]. The calculated reservoir temperatures by chemical geothermometers for the same hot spring may vary in a wide range, which is due to the errors of calculating coefficients of geothermometric equations, analyzing chemical species of samples, sampling and errors related to the geologic and thermodynamic processes of the chemical equilibriums involved in the reactions [26, 56, 78]. The ternary diagram of Na–K–Mg for the water samples from the Xi'an geothermal field indicates most geothermal waters are partially mature with temperatures less than 100 °C (Fig.11). The ^{14}C ages of the geothermal waters are from 5,500 to 34,700 a, and higher ^{14}C concentrations in the thermal waters resulted from mixing with shallow young groundwater. The hydrogeochemical data indicate that the thermal waters in the Xi'an geothermal field are likely originated from meteoric water that infiltrated to great depths along faults in the area of the Qinling Mountains [71].

3.2. Stable Isotopic Geochemistry of Oxygen and Hydrogen of Geothermal Water

Oxygen and hydrogen isotopic ratios are widely used to identify the origins of waters because different genetic waters have their typical O and H isotopic ratios. Hot spring waters are chemically correlated to chemical compositions of local rocks and deep-earth fluids since the waters are predominantly derived from meteoric water. The meteoric water line in China ($\delta D=7.9\delta^{18}O +8.2$) is comparable with the WMWL (world meteoric water line, $\delta D=8\delta^{18}O +10$). The δD and $\delta^{18}O$ data of different kinds of waters in China scatter along the WMWL (Fig.12), indicating meteoric water origin for the waters [27, 32, 71, 74, 77, 79, 80, 81, 82, 83, 84, 85]. The data of δD and $\delta^{18}O$ for spring waters in the Tianchi volcanic area are plotted nearby the WMWL (Fig.12), and chemical compositions were plotted in the corner near the Mg end in the immature region of ternary diagram of Na-K-Mg [86], indicating meteoric water origin for the spring waters and mixing of much cool water. For other geothermal systems, the shift of δD and $\delta^{18}O$ values from the WMWL can be attributed to the following processes.

The first is the isotopic latitude and altitude effect. The altitude shows an increasing tendency from the east coast to west of China. Oxygen isotopic ratios decrease linearly with increasing altitude ($\delta^{18}O$/100m=0.31‰) in the west Sichuan-east Tibet region where the altitude is from 2000 to >5000 m (i.e. $\delta^{18}O$ values vary several per mil) [84], which results in left shift of $\delta^{18}O$ for Sichuan geothermal waters and some dots on the left of WMWL (Fig.12). The $\delta^{18}O$ and δD values of the geothermal waters in the Maanping hot-spring area, Jiangxi province vary from -6.84‰ to -7.23‰ and from -44.1‰ to -48.8‰, respectively. In comparison, the values of local shallow ground waters range from -3.97‰ to -5.46‰ for $\delta^{18}O$ and from -22.3‰ to -37.6‰ for δD, respectively. The isotopic differences of oxygen and hydrogen between geothermal and local ground waters demonstrate that the geothermal waters are recharged from a higher altitude comparing shallow ground waters [63].

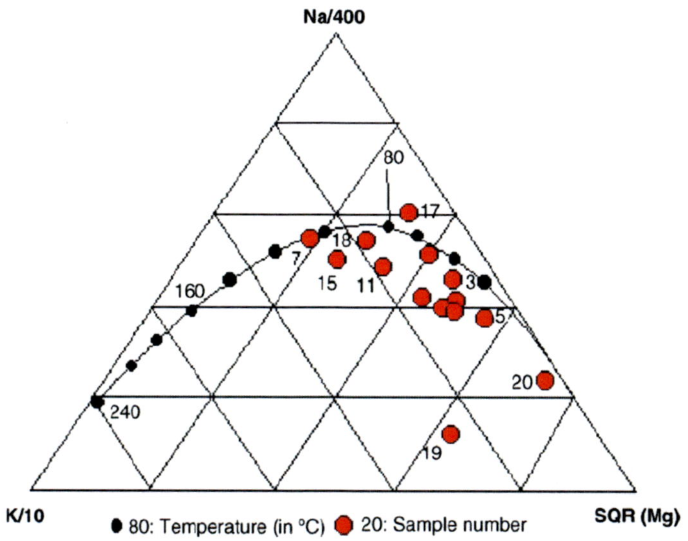

Figure 11. Na–K–Mg diagram for the water samples Xi'an geothermal field in Shaanxi [71].

The second is isotopic exchange between water and minerals. Some isotope exchange experiments on rock-fluid systems were conducted to elucidate mechanisms and rate of isotopic exchange. The effects on isotopic compositions of water depend on properties of host rocks, rate of water-rock reaction, resulting in different isotopic shift of geothermal water from meteoric water. The correlations between rates and lattice energies are quite good for mineral-H_2O systems, indicating that the more increase in rates correlated with a decrease in the electrostatic attractive lattice energy, i.e., the greater the lattice energy required to break up the crystal, the more sluggish the rates for both chemical and isotopic exchange [90]. The experiments of isotopic exchange between molecular hydrogen and hydrous minerals (epidote, kaolinite, muscovite, biotite, and hornblende) and water at temperatures between 150 and 400°C indicate that fractionation factors (α), expressed as $1000 ln\alpha$, vary linearly with temperature. Hydrogen isotope fractionations in the epidote-H_2 system increase linearly with increasing $1/T^2$ between temperatures of 150 and 400°C and become larger than 0 at temperatures of about 165°C, but values of $1000 ln\alpha$ for kaolinite-H_2 and muscovite-H_2 fractionation factors similarly decrease linearly with increasing $1/T^2$ between 200 and 275°C, and 200 and 400°C, respectively [91].

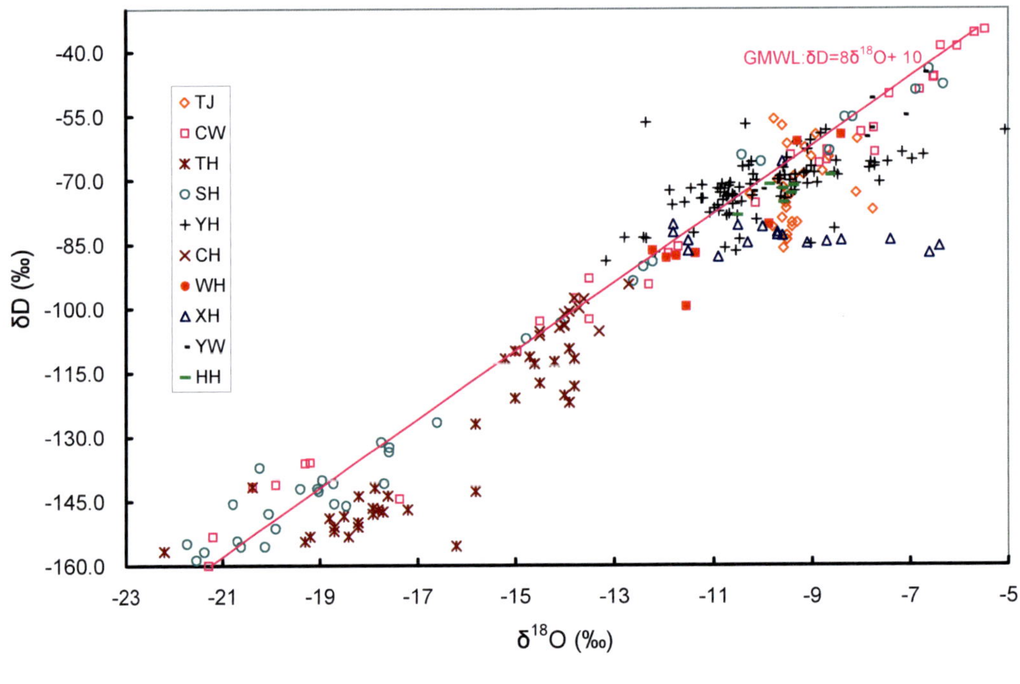

Figure 12. Diagram of δD vs $\delta^{18}O$ of geothermal and cool waters in China. GMWL: global meteoric water line ($\delta D=8\ \delta^{18}O+10$), open square (CW): some cool water (cool spring, rive and snow) [77], diamond (TJ): water in Tianjing geothermal field [74], star (TH): geothermal water in Tibet [82, 87], circle(SH): geothermal water in Sichuan province [83, 84], plus (YH): hot spring water in Yunnan province [27, 32, 80], cross (CH): hot spring water in Tianchi volcanic area [79, 88], dot (WH): spring water Wudalianchi volcanic area [85], triangle (XH): thermal water in Xi'an geothermal field [71], shot bar (YW): thermal water in Wuyi geothermal field, Zhejiang [81], and longer bar (HH): geothermal waters in Henan [89].

The third is isotopic salt effect. Feder and Taube recognized the importance of ionic hydration as an effective isotope fractionation process in aqueous solutions [92]. The isotope salt effects may change isotope fractionation factors between water and other phases by several per mil for oxygen and tens per mil for hydrogen isotopes in aqueous solutions [93, 94]. Isotope fractionation related to the hydration of ions, ion clusters, or neutral species may also play a significant role in volcanic vapors, high temperature aqueous solutions and vapors in hydrothermal systems or in surface reactions during crystallization from fluids or vapors, the latter also involving kinetic isotope effects [95, 96].

The others are vaporization, temperature, deposition, and mixture of shallow water and deep geothermal water. For example, a significant oxygen isotope shift as compared to cold ground waters is found in Xi'an geothermal field (Fig.12), which results from ^{18}O exchange between water and minerals, or cooling of much deeper geothermal waters that have already undergone strong ^{18}O exchange at higher temperatures [71], as well as isotopic salt effect.

In addition, strontium and boron isotopes and SiO_2 and Cl concentrations are also employed to trace the water origins and water-rock interaction. SiO_2 contents vary linearly with Cl contents, and Cl concentrations vary linearly with B concentrations in shallow thermal waters in the Yangbajing geothermal field. The shallow thermal waters have similar H and O isotopic ratios. These indicate mixture of shallow and deep waters in Yangbajing geothermal field [69, 97]. Mixing trends in the plots of alkali vs Cl for fluid geochemistry of Ambon Island (Indonesia) suggest the existence of a parent geothermal liquid with very high Cl content (14,000 mg/kg). The parent liquid cools through conduction and/or mixing with cold, low-salinity groundwaters. Na-Cl-HCO_3 type waters originate from dilution of the parent liquid accompanied by water-rock interaction under high P_{CO2} conditions [98].

4. Gas Geochemistry of Hot Springs in China

Gases exist ubiquitously in geothermal waters. Isotopic geochemistry of gas is major technique for tracing the origins of gases, rocks and geothermal energy. Gaseous components are usually utilized to determine temperature of geothermal reservoir. Noble gases derived from the deep-earth can provide important information about movement of the crust and mantle because these gases hardly react with other materials during migration. Helium is a sensitive tracer for both fluid migration and gas origins. $^3He/^4He$ ratios in Earth's materials have a range from about 10^{-9} to 10^{-5}. Gases of the different origins have different isotopic compositions [99, 100, 101, 102]. Most of terrestrial $^3He/^4He$ ratios can be explained in terms of mixing among three end-members: atmospheric helium with a $^3He/^4He$ ratio (R_A) of 1.4×10^{-6}, radiogenic helium (0.01 R_A) and upper mantle helium (8 R_A) [100, 103]. The characteristics of gases derived from the various tectonic regions are geochemically different [25, 26,104].

There is an intrinsic relationship among the variations of helium and carbon isotopic compositions, temperatures of the hot springs and earthquakes. The hot springs are surface manifestations of thermal fluids in the lithosphere, and the thermal fluids hold much information about the deep earth where earthquakes occurred by releasing a lot of energy. It is believed that anatectic fluids play an important role in earthquake generation by reducing friction between the fault blocks and transporting upper mantle energy [26, 105, 106]. The geochemical anomalies occur before, during and after earthquakes [107, 108]. Helium isotopic ratios are directly correlated with heat flow because heat flow is mainly derived from the crust and mantle. Helium in spring gases is also derived from the crust and mantle [25, 26, 99, 109]. Consequentially, heat flow has a close relationship with earthquakes related to both active faults and geothermal anomalies. For example, heat flow values are lower in the tectonic areas that formed in pre-Mesozoic Era and are presently not active, whereas higher heat flow values are measured in the presently active tectonic regions in China [19, 110]. Most of the earthquake epicenters occur in geothermal zones worldwide, as mentioned above. In addition, the observed carbon isotopic ratios in geothermal discharges were likely to represent frozen in compositions attained after minimum residence times of 20 ka at 400°C and 10 Ma at 300°C [55]. The origins of fluid components and heat, temperature of reservoir and geothermal resources can be, therefore, investigated by isotopic geochemical technique. Molecular compositions of spring gases were employed to trace origins of gases [57, 111] and to estimate temperature of geothermal reservoir [26, 53, 112]. Consequently, it is possible for us not only to recognize the origins of these gases, but also to look into earthquake activities based on the isotopic geochemistry of helium and carbon.

4.1. Gas Geochemistry of Geothermal Fluids

Main no-vapor gaseous components of geothermal systems in China are CO_2, N_2, and trace amount of H_2S, H_2, CH_4 (and C_2H_6, C_3H_8), NH_3, CO, O_2 and noble gases (He, Ar, Rn, Ne, Kr, Xe). The gaseous concentrations of geothermal systems are direct proportion to the temperatures of geothermal systems. The molecular and isotopic compositions of gases provide constraints on the origin of the fluids. The ratios of gas concentrations and triangle plots are, therefore, used to trace the gas origins and deep earth processes [32, 46, 77, 113, 114, 115, 116, 117].

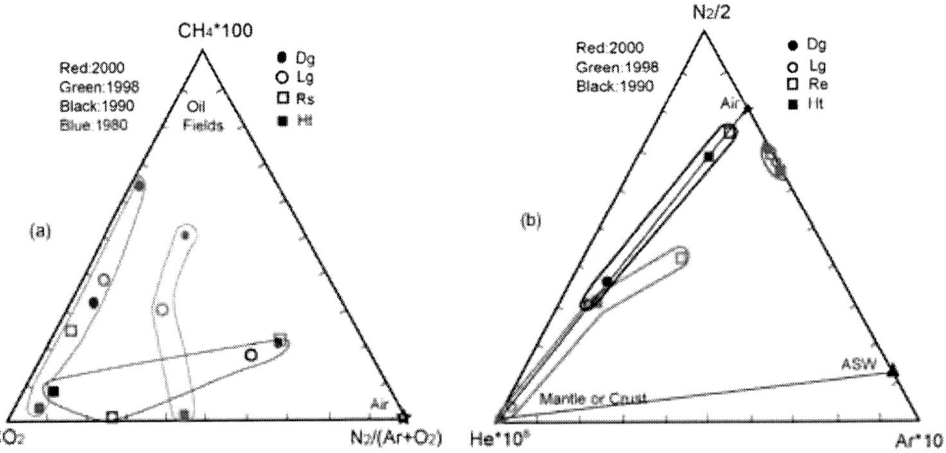

Figure 13. (a) ternary diagram of $CH_4 \times 100$-CO_2-$N_2/(Ar+O_2)$, (b) ternary diagram of He-Ar-N_2, showing variations and origins of gas components in Dg (Dagongguo), Lg (Laogongguo), Rs (Reshuitang) and Ht (Huaitaijing) hot-springs in the Rehai geothermal field with time [32].

The triangle plots of CH_4-CO_2-N_2/(Ar+O_2) and N_2-He-Ar contents (Fig.13) illustrate that concentration variations of gas components of hot-spring gases from the Rehai geothermal field, which resulted from variation of gas contributions from mantle, crust and atmosphere; and that N_2 and Ar the hot springs were largely derived from air with mixture of lithosphere gas. The variations of helium isotope ratios also indicated amount of mantle gas contribution changed temporally. Samples of 1998, collected after a big hydrothermal explosion of 1997 in the Rehai geothermal field, have the largest helium isotope ratios [32]. The helium concentrations (9.5 to 334×10^{-6}) in the hot-spring gases in the Tibetan Plateau indicate mixing of air and underground components [118]. Hydrogen concentrations of hot-spring gases in the Tianchi volcanic area range from 0.02% to 2.56%, showing deep origin for hydrogen. Methane concentrations from 1.2% to 2% and $\delta^{13}C_1$ values of -26.15‰ to -30.44‰ indicate a biogenic origin. Concentrations of CO_2 and He range from 77.8% to 97.6% and from 0.2×10^{-6} to 312×10^{-6}, respectively, indicating mantle and crustal origins [76]. Chemical compositions of hot spring gases in Hengjing, Jiangxi province are mainly consisted of CO_2 (96.7%-99.84%) as well as trace amount of N_2 (<2.01%), CH_4 (0.04%-1.86%), Ar (0.095%-0.211%), He (0.0047%-0.0234%), which indicates deep-earth origin for the gases combining with $\delta^{13}C_{CO2}$ values of -5.5‰ to -4.4‰ and $^3He/^4He$ ratios of 1.36 R_A to 2.11 R_A [119].

Gas monitoring in geothermal area where seismicity and volcanism are active is very useful for predicting volcanic and seismic activities. Wang et al. report that concentrations of He, H_2 and CH_4 seems direct proportion to earthquake activity in the Batang and Lanchang Rive earthquake zones and H_2 and He concentrations in the Lanchang Rive earthquake zone decrease with increasing the epicenter distance, and that gases are predominantly derived from crust [46]. The similar phenomena in the earthquake zones in west Sichuan are reported recently [26]. A great decrease in water pH and redox potential, and increases of dissolved CO_2 contents and $^3He/^4He$ ratios were observed in Vesuvius volcano, which supports the hypothesis of an input of magma-derived He and CO_2 into the aquifer feeding the Olivella spring by the time of the earthquake [120]. Hydrogeochemical studies in seismically active areas have also revealed their usefulness in detecting earthquake-related anomalies to be used as precursors. Isotope geochemistry of CO_2 and He in dissolved gases demonstrated that magmatic volatiles are actively transported by ground waters flowing along the main faults and fractures, and that the chemical composition of ground waters on Mt. Vesuvius greatly depends on the extent of gas-water-rock interactions, driven by the supply of acidic volcanic gases (mostly CO_2) in the aquifer [121]. Gas geochemical study indicates that probability of volcanic eruption in Wudalianchi volcanic area is very small in near future [122].

Moreover, the evolution of geothermal fluids can be investigated by studying the fluid inclusions in hydrothermal minerals. By comparing to gaseous composition ratios of air/asw (air saturated water) and combining the isotopic data of H, O and C, the relatively high $^{40}Ar/^{36}Ar$ and N_2/Ar ratios ($332<^{40}Ar/^{36}Ar<564$, $248<N_2/Ar<4197$) indicate the presence of radiogenic ^{40}Ar and excess N_2 in the fluids trapped during early-stage hydrothermal circulation from geothermal wells in the Larderello area, Italy, but the fluid inclusions formed during late-stage hydrothermal activity are characterized by low N_2/Ar and $^{40}Ar/^{36}Ar$ ratios ($309<^{40}Ar/^{36}Ar<354$, $47<N_2/Ar<66$), and similar to present-day geothermal fluids. It is indicated that fluids originated from mantle, magmatic, and meteoric fluids interacted at depths and were preserved in different proportions in the different stages of fluid evolution [117].

Molecular compositions of gases are widely applied to estimate temperatures of geothermal reservoirs [53, 78, 112]. Variations of many factors, such as the errors of calculating coefficients of geothermometric equations, analyzing chemical species of samples, sampling and errors related to the geologic and thermodynamic processes of the chemical equilibria involved in the reactions, can result in a wide range of reservoir temperatures calculated by different geothermometers for the same hot spring or well [32, 78, 123]. Among the calculated temperatures of gas geothermometers, CO_2 geothermometer usually gives higher temperature [123, 124]. The temperatures of geothermal reservoirs in the Tengchong volcanic area were estimated to be about 150-300 °C by gas geothermometers, in which H_2-geothermometer gave the highest temperatures [32]. There may be three layers of reservoirs in different depths in Tengchong volcanic area with different temperatures, about 250°C for the deep one, 240-190°C for the intermediate one and 150-195°C for the sallow one [31]. Based on the concentrations of CO_2, H_2S, H_2, CH_4, N_2 and Ar in fumaroles and wet-steam wells, otherwise, subsurface temperatures in the Hveragerdi high-temperature geothermal field, SW Iceland, were estimated to be about 240-260 °C and CO_2-geothermaometer gave the highest temperatures, indicating the calculated subsurface temperatures decrease from the northern part to the southern part of the field [123]. Temperatures of geothermal reservoirs in the Tianchi volcanic area in Changbai Mountain were estimated to be (166±9) °C with gas and isotope geothermometers [76]. In the active volcanic region of Banda arc, the main gas components in hot spring and fumaroles are H_2O, CO_2 and SO_2; CO_2 predominates over (SO_2^{2} + H_2S), and measured temperatures of volcanic fumarolic gases range from 98 to 490 °C. The calculated gas equilibrium temperature are very high, 700℃ for Wurlali and more than 1000℃ for Ili Lewotolo, reflecting magma activity in the volcanic systems [125].

Carbon isotopic geothermometers are established based on the equilibriums between carbon-bearing substances. In the case of the lower concentration CH_4 and high concentration CO_2 showing signatures of abiogenic/mantle origins [114, 126], carbon isotopic geothermometer can reflect the deep reservoir temperatures in the Rehai geothermal field because carbon isotopic equilibrium between CO_2 and CH_4 may be approached in some hot springs. The ^{13}C values of CO_2, HCO_3^- (aq) and $CO_3^=$ (aq) in hot springs of the Rehai geothermal field show, nevertheless, that isotopic equilibriums between CO_2 and HCO_3^- (aq), CO_2 and $CO_3^=$ (aq) and between DIC and travertine were not achieved, and no carbon isotope fractionation between HCO_3^- (aq) and $CO_3^=$ (aq) in the hot springs was found. Therefore, it is hard to estimate temperature of reservoir by the carbon isotope geothermometers of HCO_3^- (aq)-CO_2(g), CO_2(g)- $CO_3^=$ (aq), HCO_3^- (aq)- $CO_3^=$ (aq) and HCO_3^- (aq)- $CO_3^=$ (travertine) [32]. Furthermore, the reservoir temperatures (average of 166°C) calculated by carbon isotopic geothermometers of CO_2-CH_4 and CO_2(l)-CO_2(g) are concordant with those by chemical geothermometers in Tianchi geothermal area, and it is inferred that there is fluid-rich reservoir at the depth of about 5.5 km to which plenty of volatile from a remained magma chamber supplied [76].

4.2. Stable Isotopic Geochemistry of Helium and Carbon in Geothremal Systems

The plot of values of $^3He/^4He$ versus $\delta^{13}C_{CO2}$ can well illustrate the origins of helium and carbon dioxide combined with $CO_2/^3He$ and $^3He/^{20}Ne$ ratios [26, 115, 118, 127, 128, 129].

The values of $^3He/^4He$ versus $\delta^{13}C_{CO2}$ of the geothermal systems in China are plotted in Fig.14 together with the data of volcano fluids in Pantelleria island [130], hydrothermal gases in North Fiji basin spreading ridge, southwest Pacific [131] and Middle Ocean Ridge basalt (MORB) for comparison. Most MORBs have $^3He/^4He$ ratios between 7 and 9 R_A, a range taken as characteristic of the depleted upper mantle [100, 132]. The measured $^3He/^4He$ ratios of geothermal fluids in Tibet Plateau are higher than typical stable continental crustal values, indicating a contribution from the mantle. Both the $^3He/^4He$ ratio and $\delta^{13}C_{CO2}$ data show regional differences (diamond in Fig.14), which is attributed to the interaction of surface water with basement rocks which have different bulk compositions. The regional differences of $^3He/^4He$ ratios, however, were caused by the differential contributions of mantle-derived magma with radiogenic ^4He derived from different country rocks. The source of carbon dioxide of the hot spring gases may be marine carbonate minerals contained in sedimentary rocks in the northern sites, but is probably largely organic in the sedimentary rocks in Yangbajain geothermal waters [118, 133].

The data $^3He/^4He$ and $\delta^{13}C_{CO2}$ of hot springs in Sichuan province scatter widely (Fig.14), showing multiple origins with more contribution of mantle helium and much more biogenic CO_2 in Sichuan geothermal systems than in Tibetan ones. Hot spring gases in Hengjing, Jiangxi, are relative enriched in ^3He to those in Tibet and lack of CO_2 derived from sediment organic matter.

The values of $^3He/^4He$ and $\delta^{13}C_{CO2}$ of hot springs in Yunnan geothermal systems show mixing of crustal, mantle and atmospheric He and CO_2. It was estimated that contributions of mantle helium to hot spring gases in the Rehai geothermal field range from 27% to 52%, crustal gas contributions range from 25% to 49%, and atmospheric gas from 0 to 48% by using $^3He/^4He$ and $^4He/^{20}Ne$ ratios [128]; and estimated mantle helium from several percent to 50% by using a two-end member model [31]. $\delta^{13}C_{CO2}$ values for the Rehai geothermal field are in the range of CO_2 derived from mantle and consistent with those from other geothermal systems in magmatic environments, demonstrating CO_2 is dominantly of magmatic origin. One with the highest R/R_A value, 7.54, from Heshun in north Rehai geothermal field is comparable to the values of volcano fluids in Pantelleria Island, Italy [130] and MORB, indicating He and CO_2 are derived from mantle. Two samples with lower R/R_A ratios from Xiposhang in Rehai geothermal field are near the value of air, indicating predominantly atmospheric origin [31]. It is worth mentioned that higher $^3He/^4He$ ratios (>5 R_A) were obtained in the active period of hydrothermal explosion and small earthquakes after 1993 [45].

The $^3He/^4He$ ratios of hot springs in the Tianchi and Wudalianchi volcanic areas in northeast China scatter nearby the region of volcanic fluids in Pantelleria island [130] and North Fiji basin spreading ridge [131], but are more depleted in ^3He (Fig.14). The wide range of $^3He/^4He$ ratios indicates multiple origins for He, and hot spring gases are more enriched in ^3He in Tianchi geothermal system than in Wudalichi, i.e. more mantle He is supplied. A narrow range of $\delta^{13}C_{CO2}$ values, same as mantle $\delta^{13}C_{CO2}$ value, exhibits that CO_2 is predominantly derived from mantle [79, 122].

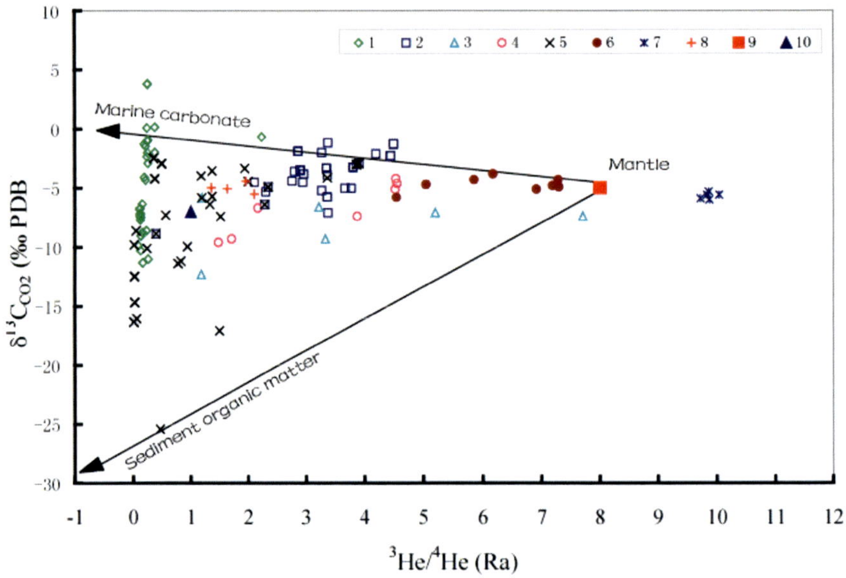

Figure 14. Diagram of $\delta^{13}C_{CO2}$ and $^3He/^4He$ of geothermal waters from China. For identification origins of helium and carbon dioxide, volcanic fluids in North Fiji basin spreading ridge and Pantelleria island are plotted. Legend: 1. geothermal waters in Tibet [118, 133, 134], 2. hot spring waters in Yunnan [32, 80, 115], 3. hot spring waters in Tianchi volcanic area [80], 4. spring water in Wudalianchi volcanic area [122], 5. geothermal waters from Sichuan [83, 135], 6. volcano fluids in Pantelleria island [130], 7. hydrothermal gases in North Fiji basin spreading ridge, southwest Pacific [131], 8. hot spring gases in Hengjing, Jiangxi [119, 136], 9. MORB, and 10. atmospheric $^3He/^4He$ ratio ($R_A = 1.39\times10^{-6}$).

The $^3He/^4He$ ratios of the geothermal gases in the geothermal system in Jiangxi, estimated temperatures from 72°C to 82°C by silica-geothermometer, range from 0.15 R_A to 0.18 R_A, suggesting input of radiogenic ^4He from crystalline rocks of the crust [64].

5. Conclusion

There are plenty of hydrothermal resources in China, and potential of hydrothermal energy is very large. More than 3,200 hydrothermal manifestations have been found in China. About 2,240 drilled wells reveal that 275 high temperature sites of hydrothermal energy, which are expected to supply a need of electric generators with a total output of 5,800 MW. More than 2,900 sites of low and intermediate temperature geothermal systems have been found, which can be utilized for heating, medicine treating, bathing, farming, etc.

Geothermal waters in China are mainly (Na, Ca)-HCO_3 type, and some (Ca, Na)-SO_4 and Na-Cl types; and the later is usually found in volcanic areas. Isotopic compositions of oxygen and hydrogen indicate the geothermal waters are derived from meteoric water, mixing small amount of magmatic volatile. Reservoir temperatures calculated with chemical geothermometers range from about 100 °C to 350 °C. Geochemical variations of geothermal fluids with time are found, which are correlated to hydrothermal eruption, earthquakes and exploitation.

No-vapor gaseous components of geothermal systems in China are predominantly CO_2, N_2, O_2, and trace amount of H_2S, H_2, CH_4 (and C_2H_6, C_3H_8), NH_3, CO as well as noble gases (Ar, He, Rn, Ne, Kr, Xe). The gaseous concentrations of geothermal systems are directly correlated with the temperatures of geothermal systems. The gases have a multiple origins of crust, mantle and atmosphere.

The geothermal areas/zones spatially overlap the seismic zones. Good correlation between geothermal-fluid geochemistry and seismic activity exists. The more amounts of mantle gases the geothermal systems contain, the higher temperature of geothermal systems and the more active the seismic zones. The deep earth fluids provide both matter and energy for geothermal fields and earthquake generation, and carry the messages of geothermal reservoir and earthquake.

More work is needed to be done for exploring geothermal energy in northwest China even though there are a lot of oil and gas in the region, but lack of surface water. More attention must be paid to protect both hydrothermal resource and environment during utilization of geothermal water.

Acknowledgment

The first author is grateful to Profs. Congqiang Liu, Wenge Zhou and Hui Zhang for their help during he works in the Institute of Geochemistry, CAS as a visiting professor, and wishes to express his thanks to the institute for supplying the nice conditions of living and working. This work was supported by Earthquake Science Foundation, China (No.B07002); the Institute of Geochemistry, CAS; and Ministry of Personnel, China (Research Fund for the scholars studied abroad).

References

[1] Fyfe, W.S., Powell, M.A., Hart, B.R., Ratanasthien, B. (1993). A Global Crisis: Energy in the Future. *Natural Resources Res.,* **2**(3):187-196.

[2] DBBI (Development Bureau of Basic Industry, State Committee of Plan and Development, P. R. China). (2000). *1999 White Book of New and Renewable Energies in China.* Beijing: China Planing Press. (in Chinese)

[3] Barbier, E. (2002). Geothermal energy technology and current status: an overview. *Renewable and Sustainable Energy Reviews,* **6**(1-2):3-65.

[4] Li, D., Zhang, L., Guo, L., Jiang, X., Zhao, Y. (2006). Alternative and renewable energies of China in the 21st century. *Natural Gas Industry,* **26** (5):1-5. (in Chinese with English abstract)

[5] Chen, M., Wang, J., Deng, X. (1994). *Geothermal Resource---the Formation Features and Potential Assess.* Beijing: Science Press.

[6] Liu, S. (2005). *Protection and Utilization of Geothermal Resources.* Beijing: Chemical Industry Press. (in Chinese)

[7] DHGE (Department of Hydrological and Geological Environment, China Bureau of Geological Survey). (2006). *Actuality of development and utilization of geothermal resources in China.* www.cgs.gov.cn/NEWS/Geology News/2006/200.(in Chinese)

[8] Wang, L., Wei, H.C. (1992). The past, the present and the future utilization of geothermal energy in Xiong county, Hebei province, Peoples Republic of China. *Geothermics,* **21**(5-6): 985-990.

[9] Kearey, P., Wei, H. (1993). Geothermal fields of China. *J. Volcan. Geother. Res.,* **56**(4): 415- 428.

[10] Wan, Z., Zhao, Y., Kang J. (2005). Forecast and evaluation of hot dry rock geothermal resource in China. *Renewable Energy,* **30**(12):1831-1846.

[11] Zhang, J., Wu, G. (2001). The present state of direct utilization of geothermal resources in the world. *Geophys. Geochem. Exploration,* **25**(2):90-94. (in Chinese with English abstract)

[12] Xin, K., Zhang, X. (1982). Geothermal resources in China. *Geothermics,* **11**(4):281-285.

[13] Huang, H.F., Goff, F. (1986). Hydrogeochemistry and reservoir model of Fuzhou geothermal field, China. *J. Volcan. Geother. Res.,* **27**(3-4):203-227.

[14] Bertani, R. (2005). World geothermal power generation in the period 2001–2005. *Geothermics,* **34**(6):651-690.

[15] Chen, M., Wang, J. (1994). Review and prospect on geothermal studies in China. *Acta Geophy. Sinica,* **37**(Sup.):320-338. (in Chinese with English abstract)

[16] Huang, S., Hu, S., Ma, L. (1986). *Volcano, Hot Spring and Geothermal Energy.* Beijing: Geological Press. (in Chinese)

[17] Huang, S. (1993). *Hot Spring Resources in China: Explanation of the Distribution Map of Hot Springs in China(1:6 Million).* Beijing: China Cartographic Publishing House. (in Chinese)

[18] Qiu, N., Hu, S., He, L. (2004). *Theory and Application of Geothermal System in Sediment Basin.* Beijing: Petroleum Industry Press.

[19] Deng, Q., Zhang, P., Ran, Y., Yang, X., Min, W., Chu, Q. (2002). Features of active tectonics in China. *Science in China (D),* **32**(12): 1020-1030.

[20] Chen, M., W ang, J., Deng, X. (1995). Advances in geothermics in China. *Earth Sci.-J. China Univ. Geosci.,* **20**(4):368-372. (in Chinese with English abstract)

[21] Chen, M., W ang, J., Deng, X. (1996). The map of geothermal system types in China and its bfief explanation. *Sci. Geol. Sinica,* **31**(2):114-123.

[22] Muffler, L.J.P. (1976). Tectonic and hydrologic control of the nature and distribution of geothermal resources. In: *Proc. Second UN Symposium on the Development and Use of Geothermal Rcsources,* san Francisco, 499-507.

[23] Henley, R.W., Ellis, A.J. (1983). Geothermal systems ancient and modern: a geochemical review. *Earth-Sci. Rev.,* **19**(1):1-50.

[24] Liu, S., Li, B., Zheng, K. (2005). *Workshop of Conference on Sustainable Development of Chinese Geothermal Industry.* Beijing: Chemical Industry Press. (in Chinese)

[25] Du, J., Xu, Y., Sun, M. (1998). $^3He/^4He$ and heat flow in oil and gas-bearing basins, China's mainland. *Chinese J. Geophy.,* **14** (2):239-247.

[26] Du, J., Cheng, W., Zhang, Y., Jian, C., Guan, Z. (2006). Helium and carbon isotopic compositions of thermal springs in earthquake zone of Sichuan, Southwestern China. *J. Asia Earth Sci.,* **26**: 533-539.

[27] Liao, Z., Zhao, P. (1999). *Geothermal Zones in Yunnan and Tibet.* Beijing: Science Press, (in Chinese).

[28] Francheteau, J., Jaupart, C., Shen, X., Kang, W., Lee, D., Bai, J., Wei, H., Deng, H. (1984). High heat flow in southern Tibet. *Nature,* **307**: 32-36.

[29] Huang, J., Zhao, D., Zheng, S. (2002). Lithospheric structure and its relationship to seismic and volcanic activity in southwest China. *J. Geophys. Res.* **107**(B10): 2255.

[30] Tong, W., Zhang, M., Zhang, Z., Liao, Z., You, M., Zhu, M., Guo, G., Liu, S. (1981). *Geothermal Energy Beneath Xizang (Tibetan) Plateau*. Princeton, N.J.: Science Press.

[31] Shangguan, Z., Bai, C., Sun, M. (2000). Character of magma gas release derived from mantle, Rehai, Tengchong geothermal area. *Science in China (D),* **30**(4): 407-414.

[32] Du, J., Liu, C., Fu, B., Zhang, Y., Wang, C., Wang, H., Sun, Z. (2005). Variations of geothermometry and chemical-isotopic compositions of hot spring fluids in the Rehai geothermal field, Southwest China. *J. Volcan. Geotherm. Res.,* **142**(3-4): 243-261.

[33] Wang, C-Y., Huangfu, G. (2004). Crustal structure in Tengchong Volcano-Geothermal Area, western Yunnan, China. *Tectonophysics,* **380**(1-2):69-87.

[34] Chen, M. (1992). A new map of hot springs in China and its explanation. *Sci. Geol. Sinica, (Sup.)*: 322-332. (in Chinese with English abstract)

[35] Xiong, S., Jin, D., Sun, Zou, Y., Fan, X., Du, X. (1991). Some characteristics of deep structure of the Zhangzhou geothermal field and its neighbourhood in the Fujian province. *Acta Geophy. Sinica,* **34**(1):55-63.

[36] Hu, S., Wang, J. (2000). Heat-flow, deep temperature and thermal structure across the orogenic belts in Southeast China. *J. Geodynamics,* **30**: 461-473.

[37] Fu, Z. (2000). The large-scale distribution of great shallow intraplate earthquakes in mainland China and adjacent areas and seismic coupling along plate margins. *J. Asian Earth Sci.,* **18**:41-46.

[38] Lin, C.W., Lee, Y.L., Huang, M.L., Lai, W.C., Yuan, B.D., Huang, C.Y. (2004). Characteristics of surface ruptures associated with the Chi-Chi earthquake of September 21, 1999. *Engin. Geol.,* **71**(1-2):13-30.

[39] Wang, J., Huang, S. (1990). Outline of geothermal resources in Taiwan Island. *J. Geomechanics*, (**2**):16-27. (in Chinese)

[40] Huang, J., Zhao, D. (2004). Crustal heterogeneity and seismotectonics of the region around Beijing, China. *Tectonophysics,* **385**:159–180.

[41] Qi, C., Zhao, D.，Chen, Y.，Chen, Q., Wang, B．(2006). 3-D P and S wave velocity structures and their relationship to strong earthquakes in the Chinese capital region. *Chinese J. Geophy.,* **49**(3): 805-815. (in Chinese with English abstract)

[42] Brander, J.L. (1974). Evidence for two possible relationships between observable surface deformation and geothermal activity. *Geothermics,* **3**(1):17-20.

[43] Barnes, Ivan, Downes, C.J., Hulston, J.R. (1978). Warm springs, South Island, New Zealand and their potential to yield laumontite. *Am. J. Sci.,* **278**: 1412--1427.

[44] Dando, P.R., Hughes, J. A., Leahy, Y., Taylor, L. J., Zivanovic, S. (1995). Earthquakes increase hydrothermal venting and nutrient inputs into the Aegean. *Continental Shelf Res.,* **15**(6): 655-662.

[45] Shangguan, Z., Zhao, C., Li, H., Gao Q., Sun M. (2005). Evolution of hydrothermal explosions at Rehai geothermal field, Tengchong volcanic region, China. *Geothermics.,* **34**: 518–526.

[46] Wang, X., Cheng, J., Xu, S., Yang, H., Xue, X., Wang, Y. (1992). Gas geochemistry of warm springs in earthquake areas. *Science in China (B),* **8**:849-854. (in Chinese).

[47] Kyung, J.B., Oike, K., Hori, T. (1996). Temporal variations in seismic and volcanic activity and relationship with stress fields in East Asia. *Tectonophysics,* **267**: 331-342.

[48] Tong, Y., Mao, J., Tong, M. (2002). The forming condition of underground heat flow and its relation to earthquake. *South China J. Seismol.,* **22**(12):53-58. (in Chinese with English abstract)

[49] Wang, J., Lin, Y., Liu, C., Liu, W. (2000). Evidence of H, O stable isotope and gas components for the formation of warm springs in south of Zhangjiakou district. *Hydrogeol. Engin. Geol.,* **4**:30-33. (in Chinese).

[50] Jiang, H., Wang, Z., Lin, Y., Tang, M. (2002).The dynamic statistical characteristics between earthquake, ground temperature and precipitation in China. *Chin. J. Geophy.,* **45**(Suppl): 144- 150. (in Chinese with English abstract)

[51] Tong, W., Zhang, M. (1989). *Tengchong Geothermics*. Beijing: Science Press. (in Chinese)

[52] Fournier, R.O., Truesdell, A.H. (1973). An empirical Na-K-Ca geothermometer for natural waters. *Geochim. Cosmochim. Acta,* **37**: 1255-1273.

[53] D'Amore, F., Panichi, C. (1980). Evaluation of deep temperature of hydrothermal systems by a new gas geothermaometr. *Geochim. Cosmochim. Acta,* **44**: 549-556.

[54] D'Amore, F., Fancelli, R., Caboi, R. (1987). Observations on the application of chemical geothermometers to some hydrothermal systems in Sardinia. *Geothermics,* **16**(3):271-282.

[55] Giggenbach, W.F. (1982). Carbon-13 exchange between CO_2 and CH_4 under geothermal conditions. *Geochim. Cosmochim. Acta,* **46**: 159-165.

[56] Verma, S.P., Santoyo, E. (1997). New improved equations Na/K, Na/Li and SiO_2 geothermamometers by outlier detection and rejection. *J. Volc. Geotherm. Res.* **79**: 9-23.

[57] Goff, F., Janik, C.J. (2002). Gas geochemistry of the Valles caldera region, New Mexico and comparisons with gases at Yellowstone, Long Valley and other geothermal system. *J. Volcan. Geotherm. Res.,* **116**: 299-323.

[58] Can, I. (2002). A new improved Na/K geothemometer by artificial neural networks. *Geothermics,* **31**: 751-760.

[59] Bin, D. (2005). Utilization and exploitation of geothermal energy in Taiwan. *Geotherm. Energy, (***4***)*:13-20.

[60] Liang, C. (1993). Geochemical characteristics of Xinzhou geothermal field, Yangdong county, Guangdong province. *Guangdong Geol.,* **8**(3): 55-62. (in Chinese with English abstract)

[61] Luan, G., Liu, H., Liu, D., Wang, W. (2002). Geothermal attributes and characteristics of warm springs in Shandong Peninsula. *Acta Geosci. Sinica,* **23**(1):79-84. (in Chinese with English abstract)

[62] Feng, Y. (2004). The formation of geothermal groundwater in the Longmen geothermal field. *J. Jiaozuo Inst. Tech. (Natural Sci.),* **23**(6): 443-446. (in Chinese with English abstract)

[63] Sun, Z.X., Gao, B., Zhang, Z.S. (2006). Geochemistry of geothermal fluids from the Maanping Hotspring Area, Jiangxi Province, SE-China. *Geochim. Cosmochim. Acta,* **70**, (18): 630.

[64] Hazaea, M., Hu, K., Mattash, M, Chen, B., Al-Jabali Aref, M.O., Abdussala, M. (2004). Geochemical survey of thermal system in Northeast China and their geothermometric characters. *J. Geosci. Res. NE Asia,***7**(2):181-18.

[65] Zhang, X., Li, G. (2006).The geologic and geochemical characteristics of the Changbai Mountain geothermal field. *Jilin Geol.,* **25**(1):25-30. (in Chinese with English abstract)

[66] Liu, J., Pan, X., Yang, Y., Wang, X., Liu, Z., Zhang, L., Zhen, K. (2002). Long-term geochemistry changes of geothermal water from a geothermal well in the urban geothermal field. *Geoscience,* **16**(3):318-321. (in Chinese with English abstract)

[67] Wang, W., Zhang, J. (1995). Geochemistry of warm springs in northwest of Hunan Province. *Science in China (B),* **24**(4):427-433. (in Chinese)

[68] Li, C., Pen, T. (2001). Resources distribution of ground warm-water in Hunan province and its perspective. *Hunan Geol.,* **20**(4): 272-276. (in Chinese with English abstract)

[69] Zhao, P., Jin, J., Zhang, H., Duo, j., Liang, T. (1998). Chemical composition of thermal water in the Yangbajing geothermal field, Tibet. *Acta Geol. Sinica,***33**(1):63-71.

[70] Xu, B., Zhang, Q., Zhou, S. (1999). Isotopic geochemistry of geothermal water in Wuyi Basin, Zhejiang Province. *Acta Geoscientia Sinica, --Bull. Chinese Acad. Geol. Sci.,* **20**(4):357-362. (in Chinese with English abstract)

[71] Qin, D., Turner, J.V., Pang, Z. (2005). Hydrogeochemistry and groundwater circulation in the Xi'an geothermal field, China. *Geothermics,* **34**: 471–494.

[72] Sun, R., Du, J. (1998). The Hydro-geochemical background of the Wudalianchi volcanic area. *Bul. Mineral. Petrol. Geochem,.* **17**(3): 150-155. (in Chinese with English abstract)

[73] Zhang, Z., Liu, S., Zhao, F. (1987). Geochemistry of thermal waters in the Tengchong volcanic geothermal area, West Yunnan Province, China. *Geothermics,* **16** (2):169-179.

[74] Zhang, B., Wang, X., Lin, J. (2006). Isotopic characteristics of geothermal water in the Tianjing geothermal field. *West-China Exploration Engineering,* (**3**):86-88. (in Chinese)

[75] Majumdar, N., Majumdar, R.K., Mukherjee, A.L., Bhattacharya, S.K., Jani, R.A. (2005). Seasonal variations in the isotopes of oxygen and hydrogen in geothermal waters from Bakreswar and Tantloi, Eastern India: implications for groundwater characterization. *J. Asian Earth Sci.,* **25**: 269–278.

[76] Shangguan, Z., Dong, J. (1997). The source of gases released from Tianchi volcanic geothermal area Changbaishan Mountains. *Science in China (D),* **27**(4):318-324. (in Chinese)

[77] Shangguan, Z., Du, J., Zang, W. (1998). Geothermal fluid geochemistry of Tan-Lu fault belts and Jiaoliao fault block. *Science in China (D),* **28**(1):23-29. (in Chinese)

[78] Zhao, P., Armannsson, H. (1996). Gas geothermometry in selected Icelandic geothermal fields with comparative examples from Kenya. *Geothermics,* **25**(3) 307-347.

[79] Shangguan, Z., Kong, L., Sun, F., Gao, S., Xu, X. (1996). Deep-seated fluid components and stable isotopic compositions in Tianchi volcanic area, Changbaishan Mountains. *Scientia Geol. Sinica,* **32**(1):54-64.

[80] Shangguan, Z., Sun, M., Li, H. (1999). Active types of modern geothermal fluids at the Tengchong region, Yunnan province. *Seismol. Geol.,* **21**(4):436-442. (in Chinese with English abstract)

[81] Xu, B., Zhou, S., Zhang, Q., Fu, W. (1997). Hydrochemical characteristic and isotopic trace in Wuyi geothermal field of Zhejiang province. *J. Sci. Techn.,* **13**(6):373-377. (in Chinese with English abstract)

[82] Liang, T., Chen, L. (1996).Hydrogeochemical characteristics of deep reservoir in northern Yangbajing geothermal field, Tibet. *Tibetan Geol.,* (**1**):43-49. (in Chinese with English abstract)

[83] Gao, Z., Yin, G., Fan, X., Wu, H., Wang, Q., Hou, M. (2004). Distribution characteristics of geothermal resources and isotopic geochemistry of hot water in the Daocheng county, Sichuan Province. *Bull. Mineral. Petrol. Geochem.,* **23** (2): 134-139.

[84] Zhang, H., Li, X. (2004). Chemical characteristics of groundwater in the Zhangla basin, northwest Sichuan. *Earth Envir.,* **32**(3-4):39-44. (in Chinese with English abstract)

[85] Gao, Q., Li, N. (1999). Asiscussion on fluid geochemistry and origin of Tengchong and Wudalianchi volcanic areas. *Geol. Rev.,* **45**(4):345-351. (in Chinese with English abstract)

[86] Lin, Y., Gao, Q., Yu, Q. (1999). Study of chemical characteristics of geothermal fluid in Tianchi volcanic region, Chanbai Mountains. *Geol. Rev.,* **45**(Sup.):241-246. (in Chinese with English abstract)

[87] Xu, Z., Yong, Z., Sun, S. (1997). The hydrochemical features of Langjiu geothermal field in Tibet. *J. Guilin Inst. Tech.*,(**1**):64-69. (in Chinese with English abstract)

[88] Lin, Y., Gao, Q., Yu, Q. (1999). Hydrogen and oxygen stable isotopic compositions and distribution of tritium in hot water of the Changbaijulongquan spring in the Tianchi volcanic region, Changbai Mountains. *Geol. Rev.,* **45**(Sup.):236-240. (in Chinese with English abstract)

[89] Wang, X., Han, P., Liao, Z., Lin, X. (2001). The geothermal method for studying hydraulic relationship among porous medium reservoirs. *J. Hydra. Engin.*, (**8**):75-78. (in Chinese with English abstract)

[90] Cole, D.R. (2000). Isotopic exchange in mineral-fluid systems. IV. The crystal chemical controls on oxygen isotope exchange rates in carbonate-H_2O and layer silicate-H_2O systems. *Geochim. Cosmochim. Acta,* **64**(5):921–931.

[91] Vennemenn, T.W., O'Neil, J.R. (1996). Hydrogen isotope exchange reactions between hydrous minerals and molecular hydrogen: I. A new approach for the determination of hydrogen isotope fractionation at moderate temperatures. *Geochim. Cosmochim. Acta,* **60**(13): 2437-245.

[92] Feder, H.M., Taube, H. (1952). Ionic hydration: An isotopic fractionation technique. *J. Chem. Phys.* **20**:1335-1336.

[93] Horita, J., Cole, D.R., Wesolowski, D.J. (1993). The activity–composition relationship of oxygen and hydrogen isotopes in aqueous salt solutions: II. Vapor–liquid water equilibration of mixed salt solutions from 50–100 °C. *Geochim. Cosmochim. Acta,* **57**: 4703-4711.

[94] Horita, J., Cole, D. R., Wesolowski, D. J. (1995). The activity–composition relationship of oxygen and hydrogen isotopes in aqueous salt solutions: III. Vapor–liquid water equilibration of NaCl solutions to 350°C. *Geochim. Cosmochim. Acta,* **59**: 1139-1151.

[95] Driesner, T. Seward, T. M. (2000). Experimental and simulation study of salt effects and pressure/density effects on oxygen and hydrogen stable isotope liquid–vapor

fractionation for 4–5 molal NaCl and KCl aqueous solutions to 400°C. *Geochim. Cosmochim. Acta* **64**: 1773- 1784.
[96] Driesener, T., Ha, T-K., Sweward, T.M. (2000). Oxygen and hydrogen isotope fractionation by hydration complexes of Li^+, Na^+, K^+, Mg^{2+}, F^-, Cl^-, and Br^-: A theoretical study. *Geochim. Cosmochim. Acta*, **64**(17):3007–3033.
[97] Zhao, P., Kennedy, M, Do r J., Xie, E., Du, S., Shuster, D., Jin, J. (2001). Noble gases constraints on the origin and evolution of geo thermal fluids from the Yangbajain geothermal field, Tibet. *Acta Petrol. Sinica,* **17** (3): 497 – 503. (in Chinese with English abstract)
[98] Marini, L., Susangkyono, A.E. (1999). Fluid geochemistry of Ambon Island (Indonesia). *Geothermics,* **28** (2, 1):189-204.
[99] Ozima, M., Podosek, F. A. (1983). *Noble Gas Geochemistry*. Cambridge: Cambridge Univ. Press.
[100] Mamyrin, B.A., Tolstikhin, I. N. (1984). *Helium Isotopes in Nature*. Amsterdam – Oxford- New York–Tokyo: Elsevier.
[101] Du, J. (1994). Helium isotope evidence of mantle degassing in rift valley, Eastern China. *Chin. Sci. Bull.* **39**(12):1021-1024.
[102] Xu, Y., Sheng, P., Liu, W., Tao, M., Sun, M., Du, J. (1998). *Geochemistry of Noble Gases in Natural Gases*. Beijing: Science Press.
[103] Du, J. (1989). Geochemistry of light noble gases. *Geol. Geochem.*, **(4)**: 56-59. (in Chinese with English abstract)
[104] Lupton, J.E. (1983). Terrestrial inert gases: isotope tracer studies and clues to primordial components in the mantle. *Annual Rev. Earth Planet. Sci.,* **11**: 371–414.
[105] Du, J., Liu, L., Kang, C. (1997). The role of deep-crust fluids in earthquake activity. *Advance Earth Science,* **12** (5): 416-421. (in Chinese with English abstract)
[106] Miller, S.A., Ben-Zion, Y., Burg, J-P. (1999). A three-dimensional fluid-controlled earthquake model: behavior and implications. *J. Geophy. Res.,* **104** (B5): 10621-10638.
[107] Rikitake, T. (1982). *Earthquake Forecasting and Warning*. Tokyo: Center for Academic Publications.
[108] Zhang, W., Wang, J., E, X., Li, X., Wang, C., Li, Z. (1988). *The Method and Principle of Prediction Earthquake by Means of Hydrogeochemistry*. Beijing: Education Science Press. (in Chinese)
[109] Polyak, B.G., Prasolov, E.M., Cermak, V., Verkhovskiy, A.B. (1985). Isotopic composition of noble gases in geothermal fluids of the Krusne Hory Mts. Czechoslovakia, and the nature of the local geothermal anomaly. *Geochim. Cosmochim. Acta,* **49**: 695-699.
[110] Cheng, P., Gao, L. (1990). Summery of studying geothermal anomaly. In Gao, L. (Eds.), *Application and Research on Geothermics*, Beijing: Seismological Publish House, 1-5. (in Chinese)
[111] Xu, Y., Shen, P., Sun, M. (1990). Non-hydrocarbon and noble gas geochemistry of natural gases from eastern China. *Science in China (B),* **6**:645-651. (in Chinese)
[112] Arnorsson, S., Gunnlaugsson, E. (1985). New gas geothermalometers for geothermal exploration — Calibration and application. *Geochim. Cosmochim. Acta,* **49**: 1307-1325.
[113] Wang, X., Xu, S., Cheng, J., Song, M., Xue, X., Wang, Y. (1993). The characteristics of hot spring gases and helium isotopic component in Tengchong volcanic area. *Chinese Science Bulletin,* **38**(9):48-51. (in Chinese)

[114] Wang, X., Chen, J., Li, Y., Wen, Q., Sun, M., Li, C., Hu, G. (1994). Volcanic activity revealed by istope systematics of gases from hydrothermal springs in Tengchong, China. In Matsuda J. (Ed.), *Noble gas Geochemistry and Cosmochemistry*, Tokyo: Terra Scientific Publishing Co., 293-304.

[115] Dai, J., Dai, C., Song, Y., Liao, Y. (1994). Natural gas Geochemistry of warm springs and the isotopic components of C, He in some areas of China. *Science in China (B),* **24**(4): 426-433(in Chinese).

[116] Gao, B., Chen, J., Wang, X. (2004). A review of gas geochemistry of continental geothermal system. *Advance in Earth Science,* **19**(2):211-217. (in Chinese with English abstract)

[117] Dallai, L., Magro, G., Petrucci, E., Ruggieri, G. (2005). Stable isotope and noble gas isotope compositions of inclusion fluids from Larderello geothermal field (Italy): Constraints to fluid origin and mixing processes. *J. Volcan. Geotherm. Res.,* **148** 152– 164.

[118] Yokoyama, T., Nakai, S., Wakita, H. (1999). Helium and carbon isotopic compositions of hot spring gases in the Tibetan Plateau. *J. Volcan. Geochem. Res.,* **88**: 99- 107.

[119] Sun, Z.X., Gao, B., Liu, J.H. (2004). Geothermal gas geochemistry of the Hengjing hot springs area in Jiangxi province. *Geoscience,* **18**(1): 116-120. (in Chinese with English abstract)

[120] Federico, C., Aiuppa, A., Favara, R., Gurrieri, S., Valenza, M. (2004). Geochemical monitoring of ground waters (1998-2001) at Vesuvius volcano (Italy). *J. Volcan. Geotherm. Res.,* **133**:81-104.

[121] Federico, C., Aiuppa, A., Allard, P., Bellomo, S., Jean-Baptiste, P., Parello, F., Valenza, M. (2002). Magma-derived gas in flux and water-rock interactions in the volcanic aquifer of Mt. Vesuvius, Italy. *Geochim. Cosmochim. Acta,* **66**: 963-981.

[122] Du, J., Li, S., Zhao, Y., Ren, J., Sun, R., Duanmu, H. (1999). Geochemical characteristics of gases from the volcanic area Wudalianchi, Northeastern China. *Acta Geologica Sinica,* **73**(2): 103-107.

[123] Sun, Z., Armannsson, H. (2000). Gas geothermometry in the Hveragerdi high-temperature geothermal field, SW Iceland. *Chnese J. Geochem.,* **19**(4):341-348.

[124] Zhao, P., Liao, Z., Guo, G., Zhao, F. (1996). Steam quantitative analysis and its implication in the Rehai geothermal field, Tengchong. *Chinese Sci. Bull.,* **41**: 501- 505.

[125] Poorter, R.P.E, Varekamp, J.C., Sriwana, T., Van Bergen, M.J., Erfan, R.D., Suharyono, K., Wirakusumah, A.D., Vroon, P.Z. (1989). Geochemistry of hot springs and fumarolic gases from the Banda Arc. *Neth. J. Sea Res.*, **24**(2-3):323-331.

[126] Dai, J., Song, Y., Dai, C., Cheng, A., Sun, M., Liao,Y. (1995). *Abiogenic Gases and Accumulation in Eastern China*. Beijing: Science Press. (in Chinese)

[127] Taylor, B.E. (1986). Magmatic volatiles: isotopic variation of C, H, and S. *Rev. Mineral. Geochem.* **16**:185–225.

[128] Xu, S., Nakai, S., Wakita H., Wang, X., Chen, J. (1994). Helium isotopic compositions in Quaternary volcanic geothermal area near Indo-Eurasian collisional margin at Tengchong, China. In Matsuda J. (Ed.), *Noble gas Geochemistry and Cosmochemistry*, Tokyo: Terra Scientific Publishing Co., 305-313.

[129] Ren, J., Wang, X., Ouyang, Z. (2005). Mantle-derived CO_2 in hot springs of the Rehai geothermal field,Tengchong, China. *Acta Geolog. Sinica,* **79**(3): 426-431.

[130] Parello, F., Allard, P., D'Alessandro, W., Federico, C., Jean-Baptiste, P., Catani, O. (2000). Isotope geochemistry of Pantelleria volcanic fluids, Sicily Channel rift: a mantle volatile end-member for volcanism in southern Europe. *Earth Planet. Sci. Let.,* **180**: 325-339.

[131] Ishabashi, J-I., Wakita, H., Nojiri, Y., Grimaud, D., Jean-Baptiste, P., Gamo, T., Auzende, J-M., Urabe, T. (1994). Helium and carbon geochemistry of hydrothermal fluids from the North Fiji Basin spreading ridge (southwest Pacific). *Earth Planet. Sci. Let.,* **128**(3-4):183- 197.

[132] Farley, K.A., Neroda, E. (1998). Noble gases in the Earth's mantle. *Annu. Rev. Earth Planet. Sci.,* **26**:189–218.

[133] Zhao, P., Xie, E.J., Dor, J., Jin, J., Hu, X.C., Du, S.P., Yao, Z.H. (2002). Geochemical characteristics of geothermal gases and their geological implications in Tibet. *Acta Petrolog. Sinica,* **18** (4):539- 550. (in Chinese with English abstract)

[134] Zhao, P., Dor J., Liang, T., Jin, J., Zhang, H. (1998). Characteristics of gas geochemistry in the Yangbajain geothermal field, Tibet. *Chinese Sci. Bull.,* **43** (21): 1770- 1777.

[135] [135]Liu, Z., Yuan, D., He, S., Zhang, M., Zhang, J. (2000). The geochemistry of CO_2-water-carbonate rock in geothermal systems and the orign of CO_2. *Science in China (D),* **30**(2):208-214. (in Chinese)

[136] Zhang, W. (2001). The application of hydrogeochemistry methods in geothermal water source analysis, Hengjin district, south of Jiangxi Province. *Hydrogeol. Engin. Geol.,* **4**:45-48. (in Chinese).

In: Geothermal Energy Research Trends
Editor: Herman I. Ueckermann, pp. 59-85
ISBN: 1-60021-683-8

Chapter 3

A Comparative Analysis of the Geothermal Fields of Larderello and MT Amiata, Italy

Giovanni Gianelli
CNR-Istituto di Geoscienze e Georisorse, Pisa, Italy

Abstract

The Larderello and Mt Amiata geothermal fields in Tuscany are large active thermal systems. Both likely overlie young plutonic rocks that serve as the principal sources of heat. The features of the two geothermal systems are similar. *1 Structural setting.* The geothermal fields of Larderello and Mt. Amiata are located in the inner part of the Northern Apennines, characterized asthenosphere uplift and delamination of the crustal lithosphere or underplating. *2 The heat source* both at Larderello and Mt. Amiata can be ascribed to the presence of shallow igneous intrusions. *3 The heat flow data* for the area surrounding both the Larderello and Mt. Amiata geothermal fields show a comparable areal extension and similar values (up to 200-300 mW/m^2). *4 Cap rocks and reservoirs.* Both the Larderello and the Mt. Amiata fields have shallow vapor-dominated sedimentary, and deep metamorphic reservoirs. At Larderello super-heated steam is present in both reservoirs, to depth of more than 3.5 km, whereas the deep reservoir of the Mt. Amiata geothermal fields is likely water-dominated. In both fields the upper reservoir is present below the flysch units forming the cap rocks. *5 Permeability* is due to rock fracturing, even at depths of about 4 km and temperatures as high as 350°C. Pressure greater than hydrostatic and a supercritical fluid can occur in the deepest part of the geothermal fields. *6 Hydrothermal alteration and contact metamorphism.* At Larderello and Mt. Amiata there is evidence of an early contact metamorphism related with the intrusion of the granites. *6 Recharge.* The water stable isotope values of the steam discharged by the geothermal wells at Larderello indicate a meteoric origin. A geochemical regional study on the thermal waters and gases of the Mt. Amiata area indicates that the geothermal reservoirs originated from a meteoric fluid, mainly stored in a regional Mesozoic dolomite-anhydrite unit, and evolved in a Na-Cl, CO_2 gas-reach reservoir by interaction with calcite-bearing metamorphic rocks.

The high temperatures existing in correspondence of a deep seismic reflector suggest the occurrence of a deep-seated unconventional geothermal resource (UGR), which can be possibly exploited. The heat could be mined from silica-rich rocks close to a plastic state, but where fracturing can be induced by fluid overpressure and abrupt high strain rates. This geothermal resource is very important, requires a re-assessment of the geothermal resources in Italy, considering the possibility of the exploitation of the new reservoirs.

Introduction

The Larderello and Mt Amiata geothermal fields in Tuscany are large active thermal systems. Both likely overlie young plutonic rocks that serve as the principal sources of heat. While there are similarities, there are also differences, such as the type of geothermal fluid: Larderello is a fully vapor-dominated system, whereas the deep reservoir of the Mt Amiata field is water-dominated. In addition, the development and the scientific data for the two areas is not the same. It is interesting to investigate the similarities and differences between these two important resources. The extension of Mt. Amiata geothermal system is still to be defined, and both systems may have, at accessible depth, a reservoir of high enthalpy supercritical fluid.

Structural Setting

The geothermal fields of Larderello and Mt Amiata are located in the inner part of the Northern Apennine, which largely corresponds with southern Tuscany (Fig. 1). The southern Tuscany and the northern Tyrrhenian basin are characterized by a shallow Moho discontinuity (20-25 km depth), a reduced lithosphere thickness, the lack of a high-velocity LID above an asthenosphere low-velocity channel and a Curie point at approximately 10 km depth (Panza and Suhadolc, 1990; Della Vedova, 1991, and related bibliographies). This is explained with the uprise of the asthenosphere and the delamination of the crustal lithosphere according to the models proposed by Reutter et al. (1980), Serri et al. (1993) or underplating (Mongelli et al., 1998; Gianelli et al., 1997a).

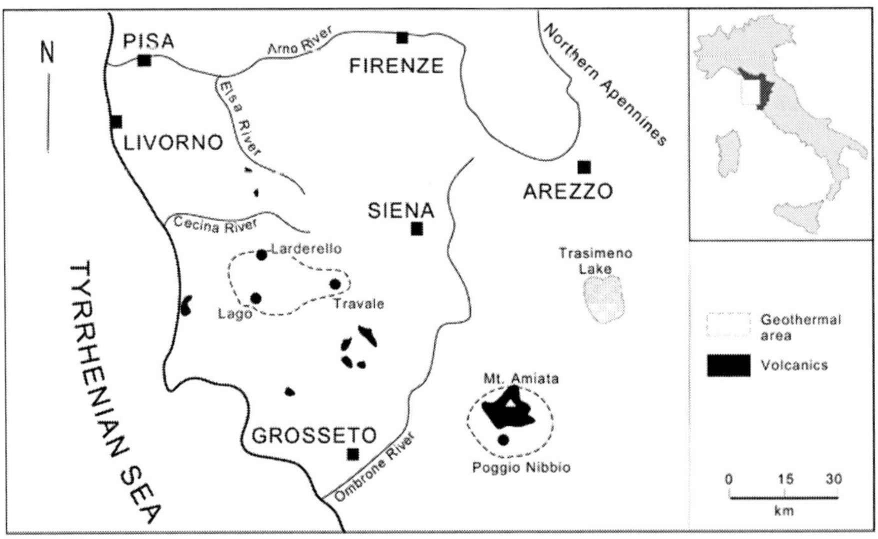

Figure 1. Location map of the geothermal fields of Larderello and Mt Amiata.

The structure of the Northern Apennines consists of an eastward-vergent compressive belt and an inner part characterized by prevailing extensional tectonics and Miocene to Quaternary magmatism (Elter et al., 1975; Reutter et al., 1980). The last phase of tectonic uplift and regional metamorphism of the Northern Apennine dates back to the Tortonian, on the basis of K-Ar and $^{40}Ar/^{39}Ar$ ages (Kligfield et al., 1986). Structural analysis carried out on Miocene-Quaternary sediments reveals a complex tectonic evolution during this time with alternating compressive and extensional tectonic events; after Pleistocene the stress field in Tuscany is characterized by the occurrence of normal faults, some of them still active (Boccaletti et al., 1985). Well breakout analysis in northern Apennine (85 wells, see Montone et al., 1999) indicates an extensional regime in the inner part of the chain (including the geothermal area of Larderello). Strike-slip faults of NE-SW direction are present and recent studies indicate that post-tectonic granite intrusions emplaced into pull-apart zones co-genetic with these strike-slip faults (Acocella, 2000).

Heat Source

Geological, petrological and geophysical data indicate that the heat source both at Larderello and Mt. Amiata can be ascribed to the presence of shallow igneous intrusions. The petrologic characteristics of the igneous intrusions found in the geothermal wells and those of the igneous rocks produced by the latest magmatic event are important data we can use to model the geothermal systems.

The magmatism in southern Tuscany and the Tyrrhenian basin (Tuscan Magmatic Province, or TMP) is characterized by a regional eastward migration with time, from approximately 7 to 0.2 Ma. There is also a broad chemical evolution of the magmatism from west to east (Boccaletti et al., 1996, and references therein). The petrologic and isotopic ages data are summarized by Serri et al. (1993). In the geothermal field of Larderello and in the surrounding areas volcanic and intrusive rocks originated by anatexis are present both in outcrops or in geothermal or mining wells. The ages range from 4.7 to 1.3 Ma (Serri et al., 1993; Villa and Puxeddu, 1994; Gianelli and Laurenzi, 2001, and related bibliographies). The rock types are peraluminous rhyolites and granites. A different composition characterizes the young (0.3-0.2 Ma) Mt. Amiata volcano, where the rocks are trachytes, trachylatites and olivine-latites.

The acidic rocks are characterized by Al-rich primary minerals such as muscovite, cordierite and andalusite. Other characteristics are high $^{87}Sr/^{86}Sr$ (0.711-0.725) values, clearly indicating a crustal origin. This is also confirmed by high $\delta^{18}O$ values, in the range +11 to +15 ‰ (Turi and Taylor, 1976; Barberi et al., 1971). Xenoliths of metamorphic rocks are frequent in both intrusive and effusive rocks. Most of them are micaschist and gneiss which underwent K-metasomatism. The melting of Hercynian rocks is shown by the U-Pb age of 245 Ma of zircon xenocrysts in the San Vincenzo volcanics, west of Larderello. Mafic enclaves are present in many granite bodies and mafic lavas flows erupted from Mt Amiata volcano. Both types have been referred to mafic, K-rich, mantle derived magmas and their occurrence is an evidence of a mixing between acidic and more mafic magmas (Poli et al., 1989; Van Bergen et al., 1983).

The chemical (Harker's) diagrams of Figure 2 reports the intrusive rocks from the geothermal wells of Larderello, which is fully consistent with the evolution trends of the granitic rocks of the region.

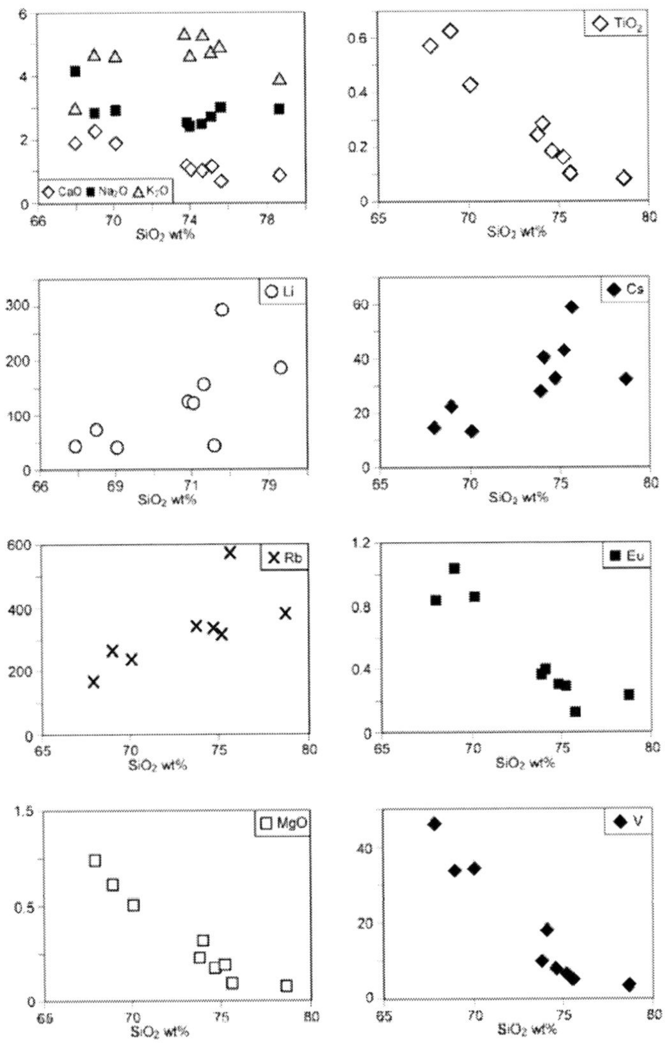

Figure 2. Harker's diagrams of Larderello granites. Values in wt% for CaO, K_2O, Na_2O, TiO_2, MgO, ppm for the others.

The acidic intrusions at Larderello (LAR granites) range in composition from leucogranite to monzogranite. Magmatic muscovites and biotites are F-rich, with values of 1-2 wt%, indicating the presence of volatiles during the magma crystallization (Ruggieri and Gianelli, 1995; Gianelli et al., 2001; Dini et al., 2004). The finding of boron-silicates (tourmaline, dumortierite) in the leucogranites is another evidence supporting this conclusion. Isotopic Sr-Nd data (Dini et al., 2004) shows that the granites cored in several geothermal wells of the Larderello –Travale field cannot be attributed to a single, isotopically homogeneous, intrusive body. Their different Sr and Nd isotopic composition suggest at least two main crustal sources, characterized by distinct $\varepsilon_{Nd}(t)$ values, activated at different time in the root zone of the geothermal field. The source with lower $\varepsilon_{Nd}(t)$ value (about –10.5) was activated earlier producing the oldest group of granites (about 3.8-2.5 Ma), then the source with higher $\varepsilon_{Nd}(t)$ value (about –8.9) was successively activated and the younger group of granites was emplaced (about 2.3-1.0 Ma). The LAR granitic system can be described as a magmatic complex built up, during a time span of about 2.8 My, by the multiple intrusions of anatectic magmas, derived by dehydration melting of different, fertile, crustal layers. At Larderello-Travale the granite samples studied did not record any significant petrographic, geochemical and isotopic features helping us to detect the contribution of mantle-derived magmas. Studies on He isotopes of both present fluids and fluid inclusions of hydrothermal minerals (Magro et al., 2003), and the finding of a mafic enclave in one well at Mt Amiata, suggest the presence of mantle-derived gas in the geothermal fluids. All the analyzed granites originated in the shallow crust under low-pressure conditions. The most differentiated terms are also characterized by very low solidus temperatures. This is in agreement with models assuming a source mainly from anatexis of pelitic rocks (Serri et al., 1993), with only a minor contribution from a mantle-derived magma. Specifically, for the granitites found in the geothermal drill holes, the micaschists and gneisses of Paleozoic or older age are very likely the source of the intrusions, and their melting has been enhanced by the enormous thermal anomaly, even at shallow depth (5-8 km), that means not far from the crustal level of emplacement of the granites.

In the LAR granitoids F, B and Cl were important components. While F mostly concentrated in the magma and was a component of micas, apatites and titanites, B and Cl exsolved from magma and entered as components of a fluid which actively participated in the first high-temperature hydrothermal stage of the Larderello geothermal system (Ruggieri and Gianelli, 2002). The P-T emplacement conditions of the Larderello granites and the Mt Amiata magmatic chamber is approximately 550-650°C and 110-120 MPa (Gianelli et al., 1996, 1988; Dini et al, 2004). These P-T conditions have been argued from: 1) a well defined pressure constrain related to the fact that the top of the granite is found at approximately 4-4.5 km depth; 2) an assumed uplift rate of 0.2 mm/year in the last 4 Ma (Del Moro et al., 1982); 3) an estimated solidus curve for the leucocratic granites of Larderello on the basis of existing experimental data of B-, Li-, and F-rich granites (Pichavant et al., 1988). The composition of the volcanic products of the Amiata volcano have been extensively studied and are similar to that of the granites cored at Larderello (Ferrari et al., 1996, and references therein).

Heat Flow

A number of geophysical data indicate the presence of a magmatic body below the Larderello geothermal field. An important seismic reflector of the bright spot type has been interpreted assuming the presence of high pressure fluids and pockets of magma at depths ranging from approximately 3-4 km below the geothermal fields of Larderello and Mt Amiata, and 9-10 km outside the geothermal fields in southern Tuscany (Batini et al. 1983; Gianelli et al., 1997a). The well "San Pompeo 2" reached a reflector 0.25 sec t.w.t above the K horizon and with same seismic features. The well found a fluid with pressure higher than 24 MPa and temperature greater than 420°C. A spontaneous blow out of the well was accompained with the emission of a high fractured and hydrothermalized schist with fractures filled by biotite, tourmaline, quartz and plagioclase.

A gravity Bouguer negative anomaly with subcircular shape is in close correspondence with the rise of the K horizon below the Larderello geothermal field. The same feature can be found at Mt Amiata. The anomaly cannot be explained with the presence of buried low density rock in the first 4-4.5 km depth, on the basis of the drilling data. Therefore this mass deficit is ascribed to the presence of magmatic bodies at depths greater than approximately 5 km.

Seismic tomography reveals a major anomaly, 20 km wide spanning from 8 to 23 km depth, and its association with a Vp/Vs ratio suggests the presence of partial melting (Batini et al., 1995). Teleseismic travel time residuals define an analogous anomaly (Foley et al., 1992). On the basis of teleseismic data the volume of the anomalous body (partially solidified granite) is estimated to be 18,000-20,000 km^3 (Gianelli and Puxeddu, 1994).

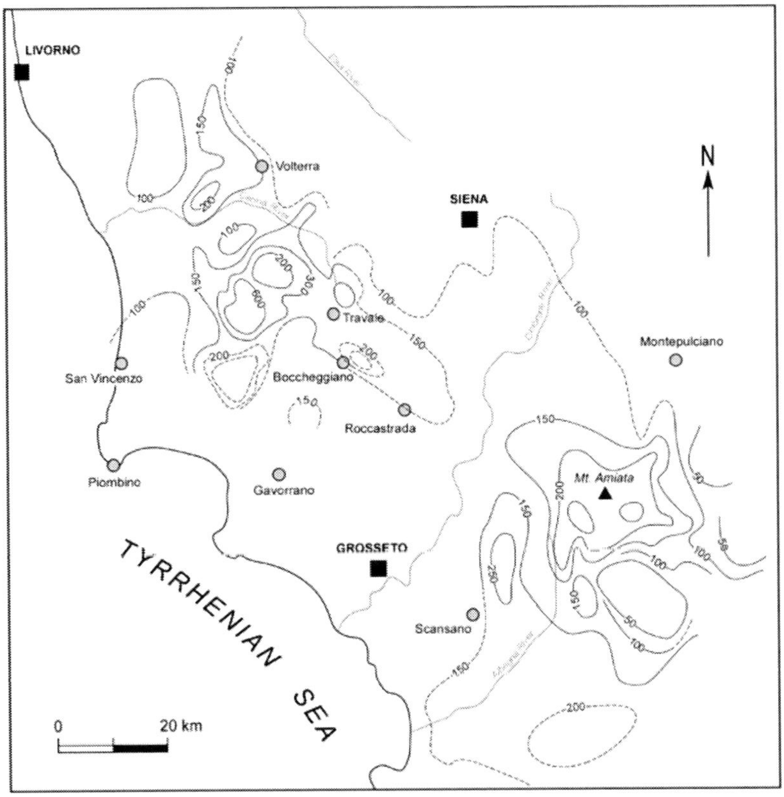

Figure 3. Heat flow map of the geothermal areas of Tuscany.

The conductivity structure of the Larderello geothermal field, defined by MT soundings, is characterized by high values in the upper and lower crust. Resistivity values greater than few hundreds of ohm-m were never been found (Fiordelisi et al., 1995; Gianelli et al., 1996). The conductivity anomaly reaches its maximum in correspondence with the seismic velocity anomaly revealed by the tomography, and extends to 12-15 km depth. Also the conductivity structure of the Larderello geothermal field is consistent with the occurrence of a partially molten granitic body. The top of the igneous intrusion should be at a depth comprised between 4 to 7 km and approximately 1-2 km below the K horizon, where the disappearance of earthquake hypocentres indicate the possible brittle-plastic transition in the crust.

Heat flow data (Fig. 3) of both the Larderello and Mt Amiata geothermal areas (Baldi et al., 1995a) are similar in size and values. Although the contoured values are not strictly the same, they are close enough to allow comparison of both the size and the general character of the anomalies. The producing areas in Larderello are overlain by values of greater than 300 mW/m^2. The thermal anomaly surrounding the Mt Amiata fields is within approximately 200 mW/m^2 contour. Moreover, it is remarkable the extension of a large anomaly south-west the area of the presently exploited geothermal fields. This thermal area includes also Hg-Sb mineralizations and thermal spring. The regional thermal anomaly surrounding the producing system at Larderello by the 150 mW/m^2 contour includes the new explored area of Monteverdi (west the old field), where steam was found in deep wells, the field Travale (Barelli et al., 1995; Baldi et al., 1995a), and to the the south-east, the mining area of Boccheggiano, interpreted as an extinct geothermal system. This wide heat flow anomaly is related to circulation of hot fluids, and part of these convecting systems can be interconnected. The geothermal system of Larderello is therefore wider than previously supposed, and the heat flow value of 150 mW/m^2 defines the area of geothermal interest with good approximation.

In conclusion, the heat flow anomalies at both Larderello and Mt Amiata are similar in size and magnitude.

Reservoir

The most obvious similarity between Larderello and the Mt Amiata geothermal fields is that both have shallow vapor-dominated reservoirs. At Larderello superheated steam is present to depth of more than 3.5 km, whereas the deep reservoir of the Mt Amiata geothermal fields is likely water-dominated (Bertini et al., 1995). The stratigraphy and tectonic units of the Larderello geothermal field are schematically represented in Figure 4. Several rock types can form the potential reservoir of the field as shown in the same figure.

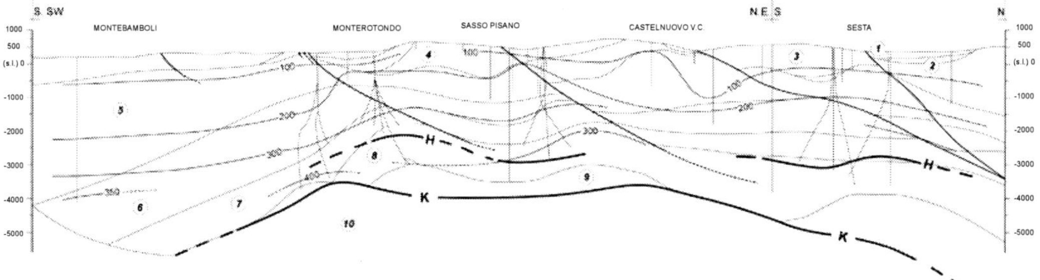

Figure 4. Geological cross section of the Larderello geothermal field (After Bertini et al., 2006, modified). 1. Neogene sediments; 2. Late Miocene sediments; 3.Flysch units (Late Jurassic-Eocene); 4. Late Triassic to Early Miocene formations; 5. Tectonic wedges, mostly dolostone, limestone, phyllite and quartzite; 6) Phyllite complex; 7. Micaschist complex; 8. Gneiss, granite, contact metamorphic and calc-silicate rocks; 9) Granites with ages of 3.8 to 2.3 Ma; 10) Quaternary igneous intrusions (size is hypothetical). K and H are seismic markers. Isogrades are in °C. Vertical and directional wells are shown with light lines, faults with heavy lines and dots (when inferred).

The upper reservoir is present below the flysch units forming the cap rocks. It consists of several rock types: sandstone, marls, radiolarites. More commonly the reservoir rocks are Mesozoic micritic limestone and anhydrite dolostone.

The upper reservoir is also formed by a tectonic complex of Paleozoic and Triassic units, mainly made up by alternating quartzite, phyllite and anhydrite dolostones (Gianelli et al., 1978; Pandeli et al., 1991). The upper reservoir rocks are highly fractured, and cataclastic rocks and tectonic breccias are present in places. This reservoir has undergone a severe lowering in production and is presently supplied with re-injected water. The lower reservoir has been extensively drilled in particular in the western (Monteverdi area) and eastern (Travale-Radicondoli area) part of the field, and it consists of metamorphic Paleozoic rocks and also Plio-Quaternary granites found in a depth range of approximately 2500-4000 m (Gianelli et al., 1978; Pandeli et al., 1994).

The reservoir fluid of the Bagnore geothermal field is a NaCl water with important carbonate content. A steam variable fraction is likely present in the reservoir. Hydrothermal breccias are present in the reservoir, and a fluid inclusion study (Ruggieri et al., 2004) depicts an evolution of the reservoir characterized by the decrease of the CO_2 (and total gas) concentrations related to boiling with gas loss and/or mixing.

Permeability, Porosity and Fracturing

Few porosity data of the reservoir rocks at Larderello are available. The porosity values of rocks of the phyllite and quartzite complex are in the range 2.4-2.9% (Cataldi et al., 1978). Bertani and Cappetti (1995) report a porosity value of 1.5% and a permeability of 10^{-11} cm^2 for the metamorphic reservoir rocks. Permeability is due to rock fracturing, even at depths of about 4 km and temperatures as high as 350°C. Flow rates of 25 kg/s have been reported from the area of Travale (Barelli et al., 1995). Strong reflectors in the metamorphic complexes have been explained with rock fracturing and presence of fluids (Batini et al., 1983; Cameli et al., 1995). The blow-out of the well "San Pompeo 2" indicates that fluids with pressure greater than hydrostatic can occur in the deepest part of the geothermal field. Permeability values for the rocks forming the potential reservoir have estimated values of 10^{-10} to 10^{-11} cm^2 for the first two km depth and 10^{-11} to 10^{-12} cm^2 for greater depth (Calore et al., 1981). There is not a systematic study of the fracture distribution in the reservoir rocks. From many geological sections available across the geothermal field we can deduce that in the metamorphic rocks most of the faults are normal and steeply dipping. Fracturing can be related with these faults. Few FMS logs and structural analysis of oriented core samples are available only from the western part of the field, and reveal the presence of extensional fractures with NW-SE strike and dipping 70 -80 NE (Baldi et al., 1995a). Gianelli and Bertini (1993) report the occurrence of an hydrothermal breccia at 1090 m depth and suggest that natural hydraulic fracturing could have been occurred during a general pressure dropping from lithostatic to hydrostatic or less than hydrostatic within the system. Hydraulic fracturing may have increased the rock permeability and in places can also be a present day mechanism of rock fracturing at Larderello, besides the regional extensional stress regime (Baldi et al., 1995a). An argument in favor of this is that, in the well "Colla 2" the reservoir at 2850 m depth consists of a fractured micaschist and modeling of VSP data suggest the presence of subhorizontal fractured layers with a thickness of tens of m (Cameli et al., 1995). Since low-angle faults (such as overthrustings) have not yet been clearly identified in the metamorphic rocks, and since most of the fractures appear to be steeply dipping, hydraulic fracturing could be a possible mechanism generating these subhorzontal structures. At Mt Amiata, deep reservoir, the occurrence of hydrothermal breccias (Ruggieri et al., 2004), lead us to hypothesize a similar model for permeability enhancement (Fig. 5).

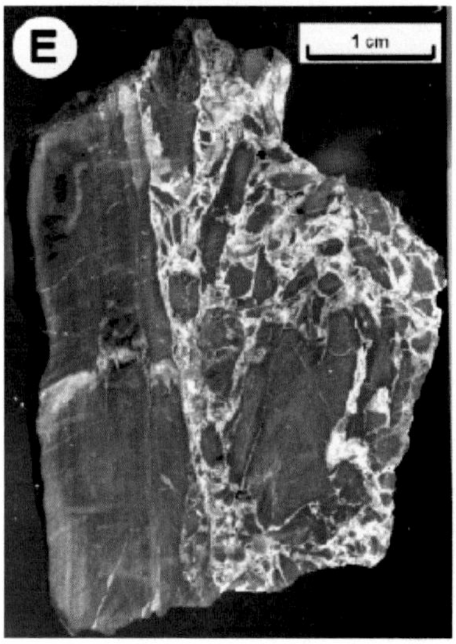

Figure 5. Hydrothermal breccia from a deep well (B3-bis, 3111 m depth b.g.l.) at Mt Amiata. The breccia elements are made up of phyllite, the cement consists of quartz, calcite, albite, adularia and minor chlorite.

Hydrothermal Alteration and Contact Metamorphism

Processes of water-rock interaction in the reservoir rocks of Larderello are indicated by the occurrence of vein minerals (Cavarretta et al., 1982; Petrucci et al., 1994). Besides this occurrence, there is evidence of an early contact metamorphism related with the intrusion of the granitic rocks (Batini et al., 1983; Cavarretta and Puxeddu, 1990; Villa and Puxeddu, 1994; Gianelli 1994). Examples are given in fig. 6.

Both geothermal fields are characterized by a boron-rich aureole related to the intrusive rocks. It is important to note the occurrence, at Larderello, in the Micaschist and Gneiss complexes, of a 2.67 Ma or younger HT-LP metamorphic event (on the basis of K-Ar ages, see Villa and Puxeddu, 1994), characterized by widespread post-tectonic biotite and the presence of corundum in equilibrium texture with K-feldspar, which allows the estimate of a temperature of about 620°C. At Larderello post-tectonic biotite, cordierite and andalusite commonly crystallize in metapelites and gneisses. The mafic hornfelses are characterized by granoblastic texture where the original oriented (lepidoblastic) amphibolite fabric is more or less disappeared. Hornblende (a green edenite, sometimes replaced by grunerite or actinolite), plagioclase (An_{10-85}), biotite, quartz and Fe-Ti-ore are the main components. This mineral assemblage indicates the attainment of the amphibolite zone and temperature possibly in excess of 600°C. The carbonatic hornfelses consist of dolomite and calcite and various silicates, such as phlogopite, wollastonite, diopside, andraditic garnet or olivine. Apart olivine the other minerals can occur also in veins. Relict wollastonite, garnet and diopsidic pyroxene indicate in places the attainment of very high temperatures (about 500°-550°C) in the past and processes of cooling of the hydrothermal system. The pressure conditions are constrained by the fact that the contact metamorphic aureole is found at 2.5-4 km depth and by the assumed uplift rate of 0.2 mm/years in the last 4 Ma (Del Moro et al., 1982).

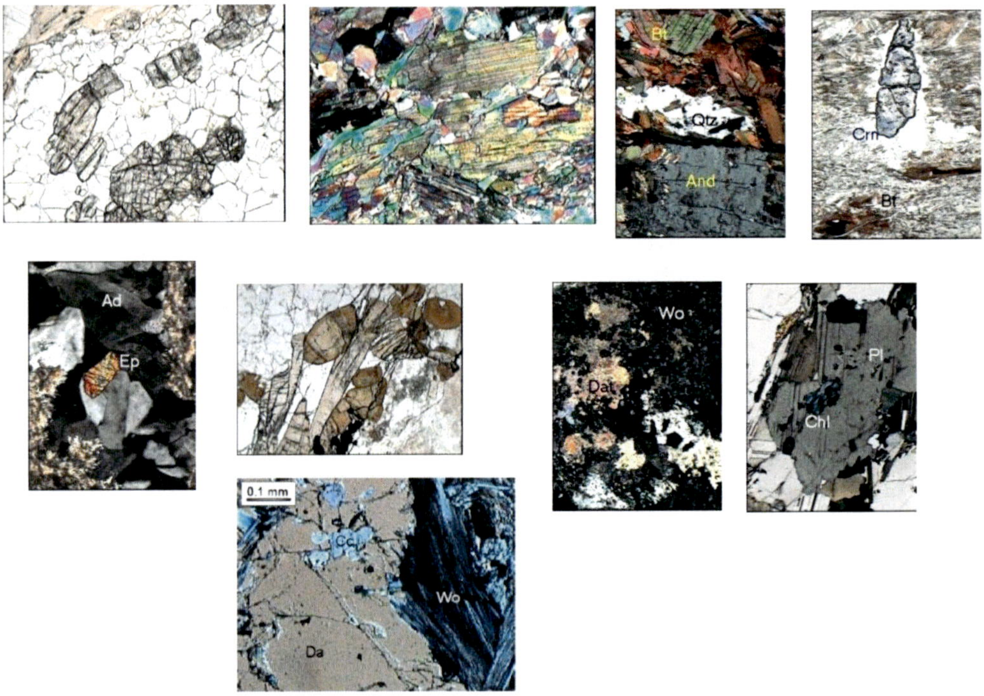

Figure 6. Examples of hydrothermal and contact metamorphism at Larderello and Mt Amiata. Upper row, from top left to right: contact metamorphic dolostone, with dolomite, calcite, diopside and phlogopite (well "Selva 4", 3374 m b.g.l., Larderello); thin levels of recrystallized anhydrite in the same dolostone; andalusite (And) in a contact metamorphic micaschist, well "Serrazzano sperimentale 1", 2244 m b.g.l., Larderello; Corundum (Crn) with a rim of feldspar in a contact metamorphic micaschis ejected fron 2900 m during the blow-out of the well "San Pompeo 1", Larderello. Lower row, from left to right: Epidote, adularia, quarts hydrothermal vein, well "Sasso 22", 1988 m b.g.l., Larderello; Quartz, tourmaline, biotite vein and breccia cement in the "Padule 2" well, 2782 m b.g.l., Larderello; Datolite and wollastonite in a calc-silicate level in the well "Sasso 22", 2600 m b.g.l.; Chlorite after biotite in a granite from a deep well (3500 approximately) at Larderello. Last row: Datolite (Da) and wollastonite (Wo) in a calc silicate level in the well "PC 34", 3277 m b.g.l. at Mt Amiata.

In the contact metamorphic rocks there is evidence of a retrograde metamorphism before the hydrothermal alteration. Epidote, serpentine minerals, actinolite and a Mg-Fe chlorite are formed during these metamorphic phase.

At Mt Amiata the contact metamorphic rocks cored so far are few. One occurrence of a calc-silicate mineral assemblage has been found so far, and consists of wollastonite, datolite and epidote, diopside-hedenbergite solid solution, and K-feldspar. Contact metamorphic dolostone with periclase has also been found in one well. An estimate of the depth and P-T conditions of the contact metamorphic rocks above the magma chamber have been made by Gianelli et al. (1988) on the basis of the xenoliths of the Mt Amiata volcanics. The values are: P = 150-250 MPa, T = 550 – 800 °C, most reasonable depth of the contact rock-magma 5-7 km.

Hydrothermal minerals at Larderello represent the filling of fractures, the cement of hydraulic breccias and, more commonly, the filling of secondary porosity originated by the dissolution of previous minerals. The latter occurrence is evident in some quartzitic phyllite, where epidote and adularia are mimetic on previous micaceous schistosity planes. The distribution of hydrothermal minerals with temperature allowed the identification of the following four "key minerals" whose first appearance indicates the attainment of well defined temperature ranges in the deep reservoir: 1) K-feldspar (T=250°C); 2) epidote (T=270°C); 3) tremolite-actinolite (T=320°C) and albite or Ca-plagioclase (T=350°C). The temperature data were based on in-hole measurements, fluid inclusion and isotope geothermometry.

At Mt Amiata the vein minerals are generally represented by quartz ± plagioclase (albite, oligoclase) ± carbonate ± chlorite ± mixed layer phyllosilicates ± K-mica ± K-feldspar (Giolito, 2005). Such assemblage has been found in a temperature range of 250-350°C. Hydrothermal assemblages made up of epidote + quartz + albite + K-felspar at approximately 350°C were also found (Gianelli et al., 1988), but are uncommon.

Fluid inclusion data indicate that the fluids present during the early "magmatic" stage at Larderello were both Li-Na-rich high-salinity fluids and aqueous-carbonic fluids with varying proportions of H_2O and CO_2. These fluids were trapped at 425-690 °C, under a nearly lithostatic pressure regime of 90-130 MPa at about 4 km depth. These pressure conditions are in agreement with the present-day depths of the contact metamorphic minerals assuming an uplift rate of 0.2 mm/years in the last 4 Ma. The Na-Li-rich fluids were probably exsolved from granites, whereas aqueous-carbonic fluids are interpreted as metamorphic fluids produced by heating (de-hydration reactions) of Paleozoic rocks during contact metamorphism. The carbonic phase of aqueous-carbonic fluids may have originated from high-temperature graphite–water interaction in the metamorphic basement (often C-rich), and/or from decarbonation reactions (Gianelli, 1985; Ruggieri and Gianelli, 2002; Cathelineau et al., 1994; Ruggieri et al., 2004).

Recharge

The $\delta^{18}O$ and δD values of the steam discharged by the geothermal wells at Larderello indicate a meteoric origin (Panichi et al., 1974). The hydrological balance of the hydrothermal system has shown that the natural recharge supply approximately 650-1300 t/h of water, corresponding to approximately 1/4 - 1/3 of the field steam production (approximately 2700 t/h); the rest of the production comes from the boiling of a deep seated fluid (Petracco and Squarci, 1976; Celati et al., 1991). The recharge mostly occurs from the south-east margin of the field (Monterotondo) and is enhanced by the exploitation, whereas in the natural state the water inflow was prevented by the vapor pressure in the reservoir (Celati et al., 1991). It is difficult to make hypothesis on the original natural conditions of the geothermal system. Indications were given from the southern part of the field, where the reservoir passed from confined to unconfined conditions. In the natural state the steam escaped from the upflow zone (the central area of the field), migrated to and condensed in the lower temperature, peripheral part of the field; after a period of exploitation the hydrothermal circulation changed and an influx of meteoric water started to flow into the reservoir (Ceccarelli et al., 1987). Another part of the field in its natural state is that of Monteverdi: here the high reservoir pressures (6-7.5MPa) indicate poor connection with the exploited part of the Larderello field (where the highest pressure are about 1-3 MPa). The temperature ranges from 270°C to more than 350°C at elevations of -1500 to -3300 m. The reservoir consists of metamorphic rocks (Baldi et al., 1995b).

A geochemical regional study on the thermal waters and gases of the Mt Amiata area lead Minissale et al. (1997) to the conclusion that the geothermal reservoirs originated from a meteoric fluid, mainly stored in a regional Mesozoic dolomite-anhydrite unit, and evolved in a Na-Cl, CO_2 gas-reach reservoir by interaction with Paleozoic units where calcite and graphite are present.

Evolution

The evolution of the P-T-X conditions at Larderello and Mt Amiata are shown by fluid inclusions and/or stable isotope data on rocks and minerals, and by their comparison with the present day fluid chemistry and temperature.

At Larderello, stable isotope data on the H_2O vapor revealed a composition of δD = -42‰ and $\delta^{18}O$ = -5.5 to 0.5‰, while the meteoric water composition is δD = -52 to -38‰ and $\delta^{18}O$ = -58.1 to -6.4‰ and led Panichi et al. (1974) to support a model of interaction between meteoric waters and the reservoir rocks. New isotopic data show that some deep wells are characterized by $\delta^{18}O$ up to 15‰ and are possibly the result of mixing between meteoric waters and magmatic fluids produced from the crystallization of a deep magma (D'Amore and Bolognesi, 1994). This hypothesis is supported by hydrogen and oxygen isotope and fluid inclusion studies. Oxygen and hydrogen isotope on silicates (Petrucci et al., 1993, Petrucci et al., 1994) indicates that: 1) The reservoir rocks in two wells (phyllite,micaschist) are depleted in heavy isotopes (values range from: $\delta^{18}O$ = 2.9 to 11.6‰, δD = -52 to -38‰) in correspondence with the occurrence of hydrothermal minerals (epidote, chlorite, quartz), and contact metamorphic minerals (biotite, cordierite and/or andalusite); 2) the computed $\delta^{18}O$ values of the fluid in equilibrium with hydrothermal and contact metamorphic minerals (quartz, muscovite, biotite and amphibole) range from 5.3 to 13.4‰, and quartz-muscovite and quartz-biotite isotope geothermometers indicate equilibrium temperatures as high as 643°C.

Fluid inclusion (F.I.) data (Belkin et al., 1985; Cathelineau et al., 1994; Ruggieri and Gianelli, 1995) reveal two main stages of hydrothermal circulation: an early stage circulation recorded only by inclusion in deep (2.9-4.0 km depth) core samples and a late hydrothermal activity testified by inclusions either in some deep core samples and in the intermediate to shallow depth (2.5-0.8 km) core samples.

Early stage fluid circulation is likely related to granite emplacement and is characterized by two types of fluid: 1) magmatic fluids, represented by H_2O-LiCl-NaCl brines with bulk salinities around 30 wt% LiCl eq.; 2) metamorphic fluids under high temperatures during contact metamorphism of C-rich and/or calcite-bearing rocks, and characterised by relatively high density CO_2-H_2O-(CH_4-N_2) vapors and liquids. These fluids are sub-contemporaneous and they were generated and trapped under high temperatures (410°-690°C), depending on the depth and location in the field, and under estimated lithostatic pressures of 90 to 140 MPa. For the metapelitic rocks of Larderello the redox conditions corresponding to the graphite-oxygen equilibrium are suggested by Cathelineau et al. (1994), on the basis of abundance of graphite-bearing phyllite in the Larderello metamorphic complexes. However, a reconnaissance isotope study on the He and CO_2 from fluid inclusions of metamorphic rocks, cored in the deepest part of the field, indicates a deep origin of CO_2 and He (Magro and Ruggieri, 1999). Carbon dioxide and other components of the gas phase can originate in the asthenosphere and reach shallow crustal levels, where they reach new equilibrium conditions with crustal rocks.

Late stage inclusions trapped a series of fluids interpreted to be meteoric waters which changed their composition and salinities through different process, as water-rock interactions, fluid boiling (salinity up to 44 wt% NaCl eq.), condensation (salinity 0-10 wt% NaCl eq.) and degassing (low density H_2O-CO_2 inclusions). Relatively high salinity brines (up to 22 wt% NaCl eq.) are possibly waters which interacted with Triassic evaporite layers.

During late stage hydrothermal activity the fluids were characterized by hydrostatic conditions (1-35 MPa) and temperatures of 200-420°C. Then the geothermal fluids evolved to present-day lower than hydrostatic pressure (usually <8 MPa) and temperature (200-350 °C). Boiling and mixing processes were very active during this decompression stage of the geothermal system (Ruggieri et al., 1999). Figure 7 portrays the evolution of the Larderello geothermal field.

The textural relationships between the hydrothermal minerals largely indicate that low-temperature minerals crystallized after the high-temperature ones. The fluid inclusions indicate the same: low-temperature secondary F.I. are always formed after the high-temperature F.I.. There is only poor evidence of F.I. decrepitation because of rising in temperature. There are rather evidences of a monotoning cooling of the system and, after its decompression, a transition from a liquid-dominated to a vapor-dominated reservoir.

At Mt Amiata fluid inclusions show that the reservoir fluid is generally characterized by a progressive decrease in gas content from an early (type I inclusions) to late-stage (type II inclusions), and then to present-day conditions. Textural and compositional data, in particular, indicate that type I inclusions, when present, record an early fluid containing appreciable amounts of gas (up to 2.8 *m*, mostly CO_2), higher than in the present-day fluid. Total dissolved solids (around 0.2 *m*) in this fluid are also higher than that in the present-day fluid. Early fluid progressively degassed during boiling. Type II inclusions record the circulation of a late, partially degassed fluid. The maximum CO_2 concentrations (up to 1.0 *m*) in these inclusions are, however, still higher than the total gas content of the present-day fluids. A further CO_2 decrease, probably due to boiling and/or mixing processes, therefore occurred before present-day. The study also shows that the CO_2 content of the past reservoir fluid, like present-day fluid, was not homogeneous.

The evolution pattern of Larderello and Mt Amiata are different, in the way that, the evidence of a past magmatic phase characterized by saline brines with high $\delta^{18}O$ values, temperatures higher than 400°C and lithostatic pressures have so far been found only at Larderello.

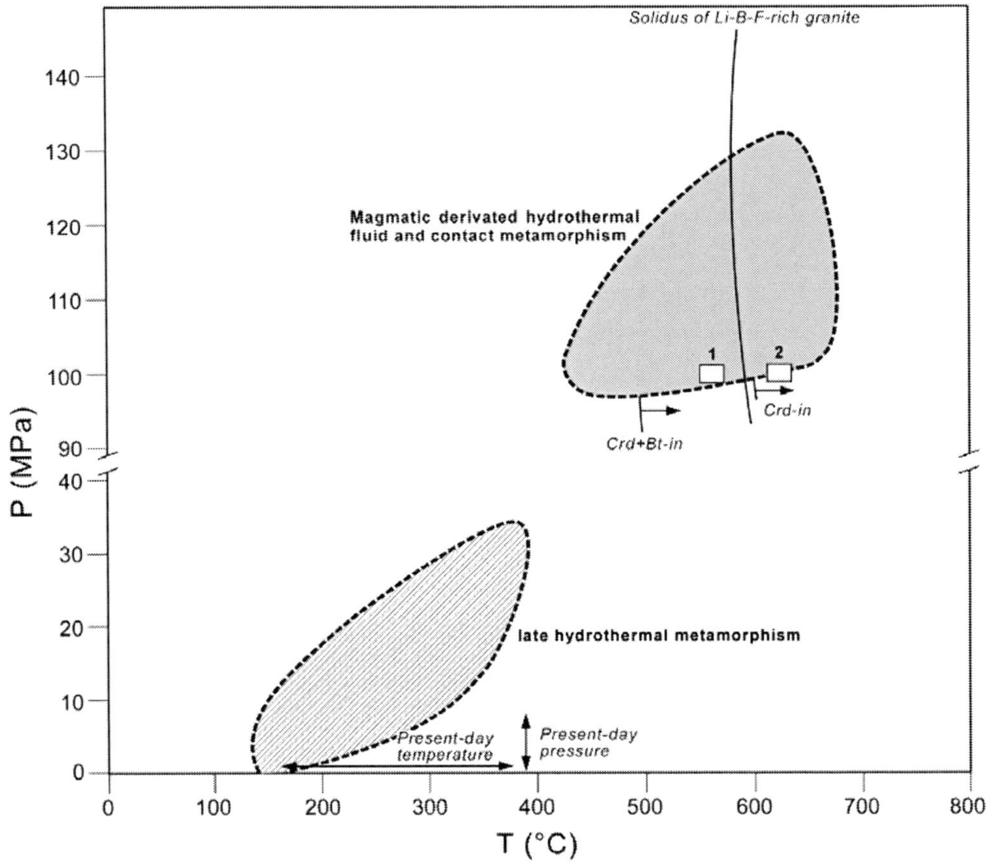

Figure 7. Evolution of the hydrothermal fluids at Larderello (after Ruggieri etal., 2002).

Deep Seated Brine

We have seen that geological and geophysical data point to the occurrence of a supercritical fluid in correspondence of the K reflection horizon at Larderello and, by analogy and in a less defined way, at Mt Amiata. But what exactly this means? How hot and how dense is this fluid? And which is the permeability (if any) in such a reservoir, made up by rocks close to the brittle-ductile transition. If we want to go deeper in this matter, we enter in a field is still largely unexplored, but some solid conjectures can be put forward, either on the physical and chemical properties of the fluid and on the characteristic of the reservoir.

A geothermal brine is a complex multi-component system, which has its proxy in the binary H_2O-NaCl system, that is not simple at all. There is a large portion of the P-T-X space were a liquid (high-salinity) coexist with a vapor (less saline), in correspondence of the depths were shallow intrusions can trigger advective heat and solute transport through the earth crust (Bodnar et al, 1985). Different equation able to compute the properties of the NaCl-H_2O solutions, and different phase diagrams have been published (among others: Chou, 1987; Bischoff and Rosenbauer, 1988; Bischoff and Pitzer, 1989; Bodnar, 1995).

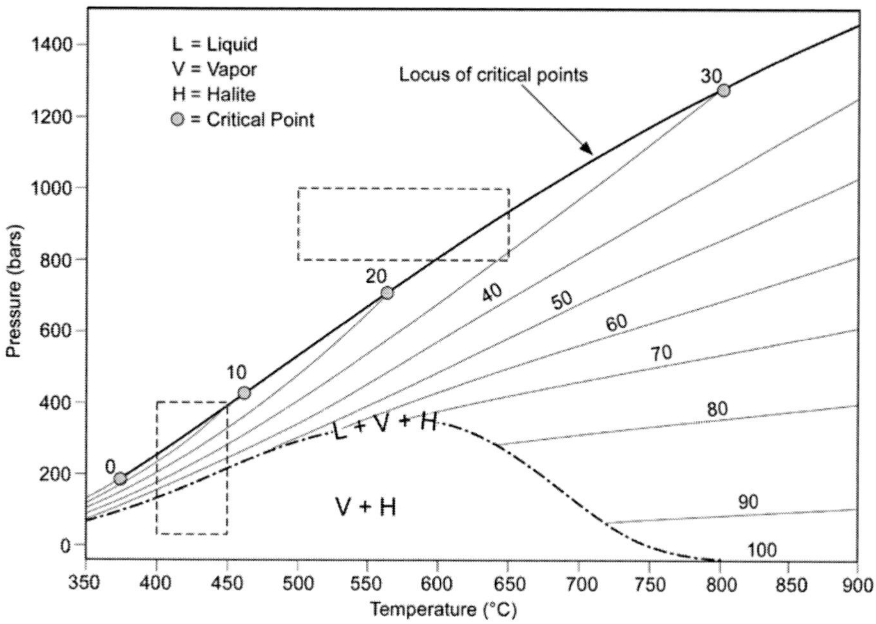

Figure 8. Phase diagram of the system H_2O-NaCl for different wt% NaCl values. L = liquid, V = vapor, H = solid halite. After Atkinson Jr. (2002). Square box = P-T conditions at the present seismic reflecting horizons K (high pressures) and H (low pressures).

Considering two different P-T phase diagrams (Bodnar 1985, Chou, 1987) it is possible to see (Fig.8) that at approximately 80-100 MPa pressure and temperature in the range of 500-650°C (the conditions of the K-horizon at Larderello), the fluid is largely in the field of a supercritical vapor-like fluid. At 400-450°C and less than 40 MPa (the possible conditions of a H-horizon, or a de-pressurized K-horizon), the fluid can be a two-phase saline brine or a supercritical fluid. Only if temperature keeps values over 400°C and pressure drops to values below hydrostatic the fluid can enter the field of the vapor + solid phase (halite). Similar conclusions can be drawn after inspection of P-T diagrams proposed by Geiger et al. (2006): at 500-650°C and 100 MPa a saline brine of 3.2 to 40 NaCl wt% concentration is a liquid + vapor or a supercritical fluid, with density in the range of approximately 600 to 800 kg/m^3 (fig. 9). Also according to these phase diagrams, a decompression of the system can drive the fluid to the vapor + halite field if pressure drops below 40 MPa. An excellent P-T-X diagram to further confirm this conclusion is that of Fournier (1987), applied to the geothermal system with a seawater recharge (Fig. 10). At the K-horizon conditions of Larderello (100 MPa and 600-650°C) a 3.2 NaCl wt% steam-rich phase coexists with a very saline fluid (with salinity of the order of 40%). Under these conditions the fluid is approximately 100-120°C lower than the minimum melting curve of the wet granite. For the Larderello granites this curve has been estimated 500-650°C by Dini et al., 2005), because of the abundance in B, F and Li. This means that, if a two-phase saline fluid (3.2 NaCl wt%) was present during the early hydrothermal stage, it could coexist with a molten rock or a crystal mush. Also from this diagram it is possible to see that at high temperatures (of the order of 400-450°C) and low pressures two type of fluids can be possible: a vapor coexisting with a solid, or a two-phase (liquid + vapor) saline brine.

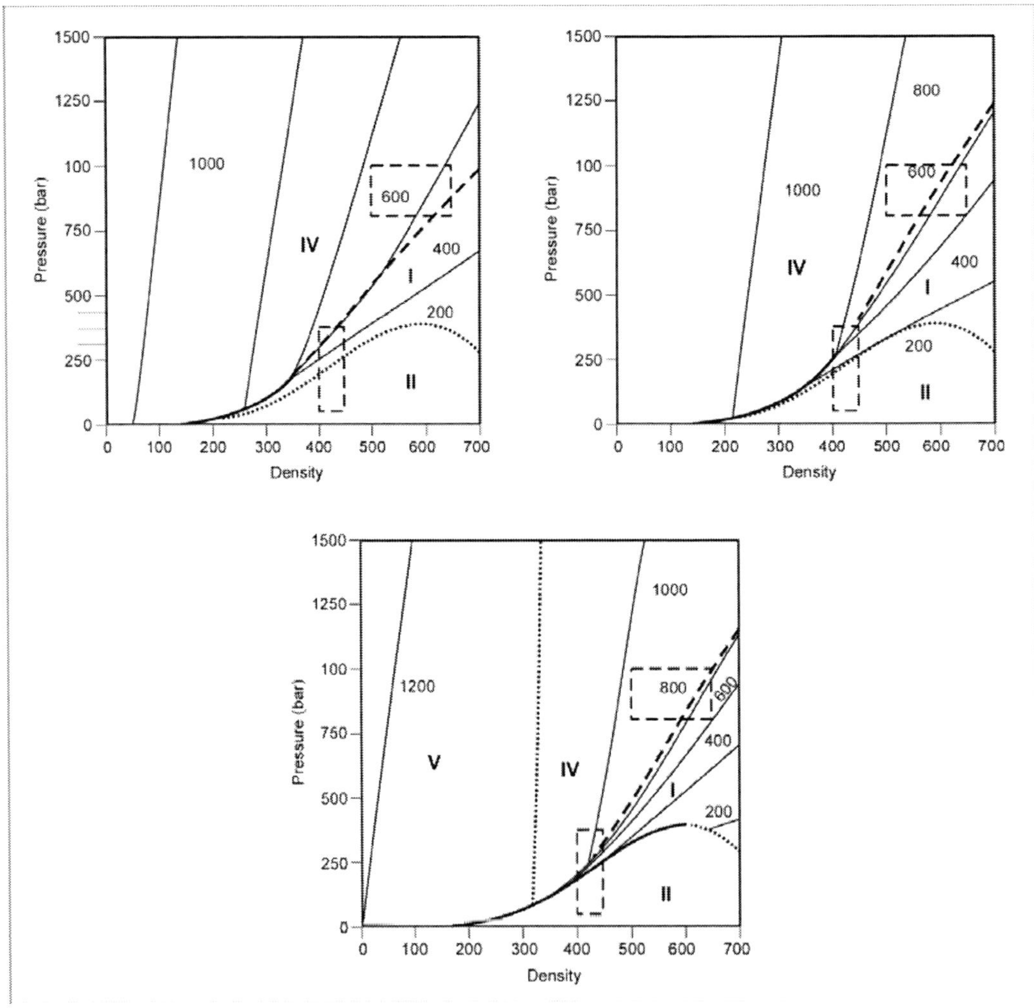

Figure 9. Phase diagram of the system H_2O-NaCl at different wt% NaCl values. A = 3.2 wt% NaCl, B = 20 wt% NaCl, C = 40wt% NaCl. I = two-phase region (L+V); II = halite + vapour region; IV = supercritical region; V = halite + L (after Geiger et al. (2006), modified. Boxes are the inferred P-T conditions at the present seismic reflecting horizons K and H.

Addition of other components can modify this scenario. In particular, for geothermal systems, it is important to consider the major component of the gas phase, the carbon dioxide. The effect of increasing the CO_2 component in a H_2O-NaCl-CO_2 system, at constant P and T, is to drive the fluid from a two-phase (NaCl-rich liquid plus H_2O+CO_2 –rich vapor) to a single phase supercritical fluid (Bowers and Helgeson, 1983). If salinity and CO2 content are in the likely range of 6-20 wt% NaCl and 0-30% CO_2 respectively, a pressure and temperature decrease to say 20-35 MPa and 400-450°C can drive the fluid to liquid+vapor, rather than an halite + vapor phase. This seems strongly suggested by the experiments of Anovitz et al. (2004), at 50 MPa and 500°C.

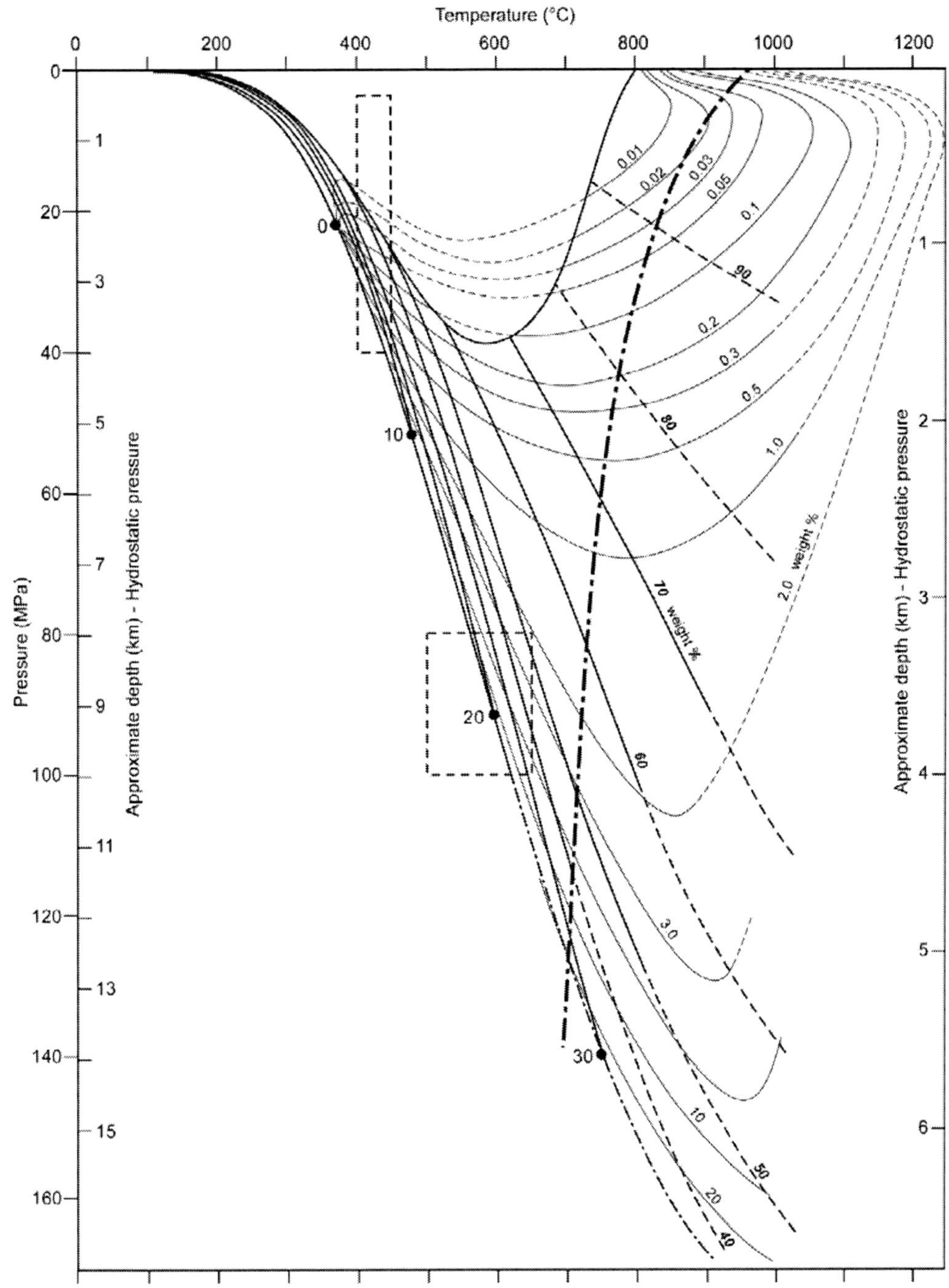

Figure 10. Phase diagram of the system H_2O-NaCl, after Fournier (1987), with the P-T conditions of the present seismic reflecting horizons K and H at Larderello (boxes). The two-phase region (L+V) is represented with lines of equal concentration in the vapor phase (light solid-dashed lines) and coexisting L+V lines (heavy solid-dashed lines). The solidus for the wet granite is represented by a dot-and-dash line.

In conclusion, the fluid present in the K-horizon is likely supercritical if an important gas (CO_2) component is present. Alternatively, if this fluid can be assimilated to a NaCl-H_2O binary system, then it can be a two-phase (liquid + vapor) or a supercritical fluid, with a density of 600-800 kg/m^3. These conditions are valid if the system is under lithostatic conditions. Pressures lower than lithostatic are instead consistent with the presence of a vapor coexisting with a very dense NaCl brine, or even halite + vapor. The possibility that part of the deep reservoir is under the H+V conditions is suggested by Truesdell et al. (1989) and D'Amore (1989), who argued that the occurrence of several hundreds of HCl in the steam produced by some wells at Larderello may be related to an high temperature, low pH and high chlorinity fluid, but, and even more relevant to our discussion, to an ambient without liquid water, which otherwise will scrub hydrogen chloride from the steam. D'Amore (1989) suggested the possible production from "deep, saline and hotter zones", below the exploited reservoir at that time, when wells produced on average from 2-2.5 km depth only. He also suggested that the hydrothermal alteration in the deepest part of the geothermal system was characterized by halite hydrolysis. So we can conclude that the presence of HCl in the steam at Larderello supports the hypothesis of the presence of a deep zone were solid NaCl and vapor coexist.

We have already discussed the data leading to the hypothesis that the H-horizon can be considered a part of the upper crust where the P-T conditions originally were those of the present-day K-horizon: lithostatic pressure, temperature close to the granite solidus, a saline brine near or at the supercritical conditions. The evolution to the sub-actual conditions, with nearly hydrostatic pressures and temperatures at least 200 °C lower that the original ones, are probably due to: 1) an abrupt nearly isothermal decompression, followed by a drying up of the brine, which reached the halite + vapor field; 2) intrusion of meteoric water and lowering of temperatures, giving origin to a fluid of moderate salinity, mostly in the liquid phase, boiling in places; 3) a further decrease in pressure to sub-hydrostatic values due to the lowering of the water table. This last phase was at least in part natural, because super-heated steam existed since before the industrial exploitation at Larderello. It is impossible to quantify the effect of the industrial exploitation of the dropping of the water table all over the area of Larderello and Travale.

Let us now consider the nature of the medium hosting the high temperature fluid. The possibility of advective circulation in rocks close to the brittle-ductile transition are matter of scientific debate.

Manning and Ingebritsen (1999) and Ingebritsen and Manning (1999, 2003) were able to derive a permeability-depth relationship for the continental crust, constrained by geothermal data and quantitative evaluations of fluid fluxes during metamorphic processes:

$$\log (k) = -14 - 3.2 \log z \text{ (with k in m}^2 \text{ and z in km)}$$

According to this equation permeability values around 10^{-18} m^2 are reached at approximately 12 km depth. This value is considered critical by Manning and Ingebritsen (1999) because it corresponds to the brittle-ductile transition in the normal continental crust. Below this permeability threshold the heat is transported by conduction, but advective transport of solutes is still possible, the limiting value for this process being estimate at 10^{-20} m^2. According to Ingebritsen and Manning (1999, 2003) these permeability values are in agreement with crustal models assuming advective transport of heat and solute, and external (meteoric) origin of fluid dominate, in the brittle crust, and internally generated fluids in a much less permeable medium where the heat transport is largely conductive below the ductile-brittle boundary. This ductile, metamorphic domain is not completely impervious and fluid generated by processes of devolatilization (loss of water, CO_2 etc. from the rocks) can migrate to the brittle "hydrothermal" domain. There are variations to these scheme depending on the crustal environment. For contact metamorphic environments at P-T conditions comparable with the present K-horizon at Larderello and Mt Amiata (4-7 km) a significant convective heat transport and mass transfer will occur at 600-650°C and pressures of 50-100 MPa, if permeability is of the order of 3 to 10 x10^{-16} m^2 (see fig 3, Manning and Ingebritsen, 1999). We have already noted that in both geothermal systems there is evidence of a deep-seated high-temperature hydrothermal circulation, even at sub-magmatic temperatures.

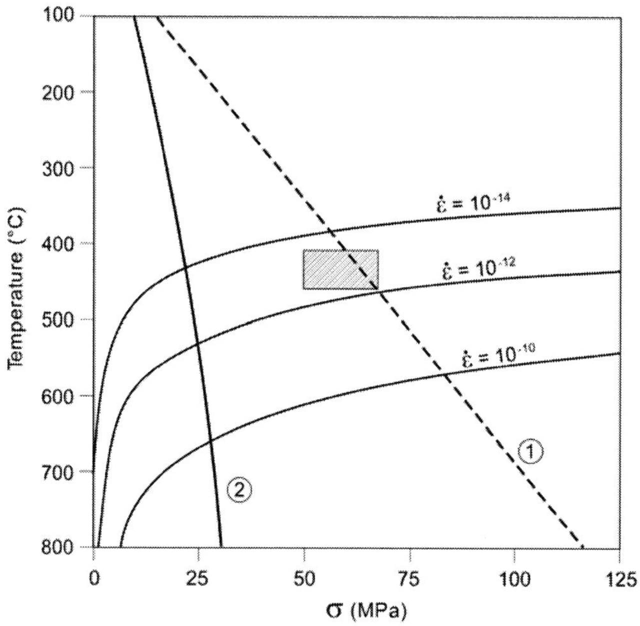

Figure 11. Effective strain vs. temperature diagram, with the conditions for brittle shear failure (straight line) and for plastic deformations (exponential curves) at different strain rates (sec^{-1}) . The box correspond to the assumed P-T conditions in the deepest part of the Larderello geothermal field (4-5 km depth, near or at the K horizon). Brittle deformation is possible if strain rates higher than 10^{-14} sec^{-1} occur.

The problem is then to explain the reasons of such a relatively permeable crust, able to act as a dynamic fluid reservoir, even in the presence of a quasi-plastic mechanical behavior of the rock. Laboratory experiments give us some partial answers to this problem. Hashida et al (2001) Tsuchiya et al. (2001) and Takahashi et al (2002) conducted high-temperature and pressure (up to 100 MPa and 600°C) autoclave experiments on granite samples and discovered that permeability is enhanced at supercritical conditions (above 400°C), but is insignificant at lower temperatures. The supercritical phase was found to be not homogeneous, having a vapor-like (below approximately 25-40 MPa at 400-600 °C) and a liquid-like (above these limits) behavior. Rock dissolution was important in the liquid-like, moderate in the vapor-like region. Besides, near the α-quartz to β-quartz transition (573-600

°C at the pressures of the experiment) the rock was able to yield open fractures through the 96 hours of the experiment run. Permeability is enhanced because of the formation of a dense microcrack network. It is interesting to note that the highest number of microcrack per unit volume occur in correspondence of 570-600°C and 40 to 100 MPa, i.e. the assumed conditions of the contact metamorphism at Larderello, and also compatible with those at the depth of the present-day K-horizon. These experiments do not give explanation of the apparent contradiction that the permeability is enhanced at higher temperatures.

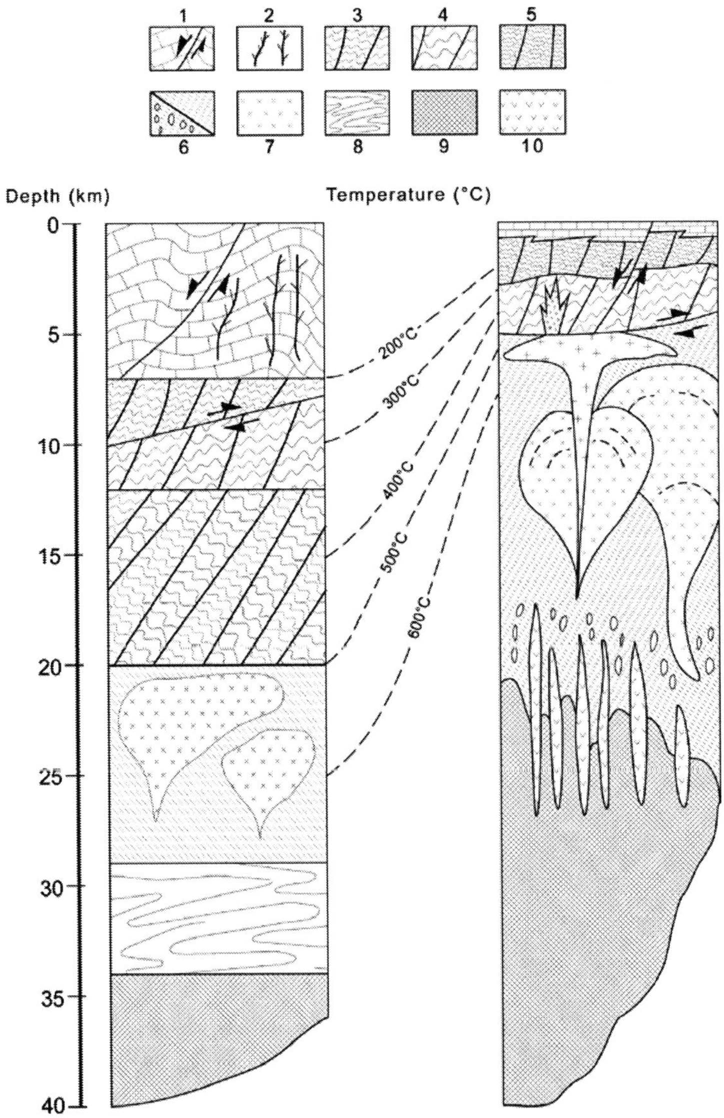

Figure 12. Hypothetical crustal section for the Tuscan geothermal areas (left) compared with the normal crust (right, after Carter and Tsenn, 1987). 1). Non-metamorphic brittle crust; 2) hydrothermal fluids; 3) very low-grade metamorphic brittle crust; 4) low-grade metamorphic crust (semi-brittle or ductile in normal crust); 5) low- to medium grade metamorphic crust (semi-brittle to ductile); 6) ductile crust of medium to high-grade, with partial melting; 7) granite or granodiorite intrusions; 8) granulite; 9) upper mantle; 10) mantle derived basaltic melts.

Hydrothermal reaction experiments were conducted on an aplite and granite samples at effective pressure of 50 MPa, variable differential stress values and temperatures up to 600°C (Morrow et al. 2001; Tenthorey and Fitz Gerard, 2006). Also in these experiments there is a much higher loss of permeability at higher rather than lower temperatures. SEM and TEM analysis (Tenthorey and Fitz Gerard, 2006) showed that there are two competing factor enhancing of hampering the permeability: microcrack opening by high pore pressures, and sealing by mineral precipitation. At low temperatures sheet silicates, whose shape is more efficient to close the pore space, are formed. At higher temperature the sheet silicates destabilize, and corundum of andalusite are formed, which have a minor effect on the closure of the pore space. The results of this experiment are also relevant for the interpretation of the K-horizon: at the present P-T conditions the rock can still be permeable and a fluid can be trapped in a high temperature silicic rock above the temperature of the muscovite break-down. It is also worth of note the evidence of minerals such as corundum and andalusite found in the H-horizon, interpreted as a reservoir which underwent previous high T and P conditions, comparable with those of the K-horizon. The α- β-quartz transition could be a further factor enhancing permeability according to Marini and Manzella (2005).

The mentioned laboratory experiments reproduce the P-T conditions of fault sealing and opening predicted by the fault-valve model of Sibson (1992). Fournier (1999) proposed a model of episodically breaching of a narrow, self-sealed zone between the brittle and the ductile layer below the zone of the earth's crust that are thermally anomalous. The breaching is due to episodes of fluid overpressure, linked to magma upsurges, whit the consequence of increased strain rate and shear failure of previously ductile rocks. Rowland and Sibson (2004) investigated the rift system of the Taupo Volcanic Zone, New Zealand, and concluded that the intensely fractured accommodation zones of the rift are the main pathways of the geothermal fluid. The convective hot fluid circulation through the live of the geothermal systems is enhanced by the continuous renewing of the vertical permeability by either tectonic increasing of differential stress or fluid overpressure.

High strain rate values are necessary to keep brittle conditions in a deep-seated geothermal reservoir with a composition where quartz is a major component (granite, granodiorite, gneiss etc), which should be normally ductile at temperatures over 450°C at a pressure of 100MPa.

Caporali et al. (2003) estimate a strain rate of 3 to 4 10^{-15} sec^{-1} for the central Apennines, on the basis of GPS data and reviewed geophysical data. At such normal values any silica-rich crustal rock of the medium to high metamorphic grade, present in the geothermal areas of Tuscany, will certainly be plastic. Higher values (10^{-12} sec^{-1}, and even less) for the geothermal areas of Tuscany have been suggested by Fournier (1991) and Gianelli (1994b). Such high values can derive from: 1) rise and emplacement of shallow magmatic intrusions, as suggested for the Mt Amiata by Acocella 2000), 2) fluid overpressures within pre-existing fractures and faults, whose orientation is favorable for their re-opening (Gianelli, 1994c).

Conclusion

1. Both fields are related to the emplacement of acidic intrusions. The areal extension of the intrusions below Larderello is more or less defined, whereas the size and depth of the intrusions below the Mt. Amiata area (including the south-west are characterized

by heat-flow anomaly), is still to be defined. The age of the Larderello intrusion is remarkably old (3.8 Ma), but younger ages (2.5 to 1.3 Ma) of other parts of the intrusion suggest a composite pluton, or a very large magma chamber slowly cooling and repeatedly fed by magma inputs from the lower crust or the upper mantle. At Mt. Amiata similar models cannot for the moment be sustained.

2. A contact metamorphic aureole is present in both fields, indicating an early phase of high temperature and interaction between the reservoir rocks and magmatic saline brines. The present occurrence of similar fluids is not demonstrated. A minor contribution of magmatic fluids is supposed both at Larderello (D'Amore and Bolognesi, 1994) mainly on the basis of geochemical data. At Larderello saline brines and/or gas are considered to exist in correspondence of the seismic reflector K, but their composition and nature are still to be understood. At Mt. Amiata the few contact metamorphic rocks, the occurrence of B-silicates (datolite) and the rare findings of granite and mafic enclave suggest a situation comparable with that of Larderello, the main difference being in a greater depth of the intrusion(s).
3. The simple "heat-pipe" conceptual model, applied Larderello (White et al., 1971; D'Amore and Truesdell, 1979; Pruess et al., 1987) does not entirely explain all the features of the two geothermal fields. A simple model assuming one or more zones of upflow of vapor from a deep-seated boiling brine and a condensate zone on top of the reservoir has been challenged by Minissale (1991), who assumed a decompression phase of the system and the occurrence of a deep supercritical fluid reservoir above an igneous intrusion and below the steam reservoir at Larderello. Tresdell (1991) supports, for Larderello, a conceptual model which assumes the drying out of an early hot-water system heated by igneous intrusions and the generation of a heat-pipe mechanism during a very recent stage of the geothermal system. Regarding Larderello, the textural relationships between the hydrothermal minerals and the cross-cutting relationships of the different sets of F.I. do not support the past occurrence of a hot dry rock stage after a drying out of previous hydrothermal circuits by increase in temperature due to the intrusion of shallow plutons. A deep (supercritical) brine is likely confined in a very hot reservoir, where rocks have a mechanical behavior transitional between the brittle and the plastic deformation. The fluid trapped in this particular environment can be released only temporally, when the temperature decreases, the strain rate increase (perhaps by hydraulic fracturing) and the state of stress is favorable (Gianelli, 1994b).

A last but not least question is about the possible exploitation of the deep-seated geothermal reservoirs (DSGR). Low-permeability but very high-temperature rocks exist both below Larderello and Mt. Amiata, and the extraction of secondary, or enhanced heat recovery from them could be possible. A pre-feasibility study of heat recovery by injection and production of water in a deep reservoir was made at Larderello (Gianelli et al., 1997b, Fig. 12). Many studies remain to be done on this matter. The main problem is to understand the mechanical behavior of the rocks at such high temperatures. If the reservoirs are, in fact, in a mechanical state transitional from brittle to plastic, the interconnection between the fractures should not be able to keep a contact between the fluid and the rock surface sufficient to allow a heat transfer. The challenge is risky, but the reward is promising: a clean, sustainable and huge resource. Of extreme interest, are the recent results from a few laboratory experiments conducted in Japan (Hashida et al., 2001; Tsuchiya et al., 2001), as mentioned above. The experiment demonstrates that at temperature and pressures likely present in correspondence of the main reflecting horizon, the DSGR can have permeability and limited mineral dissolution-precipitation, and could be suitable for exploitation.

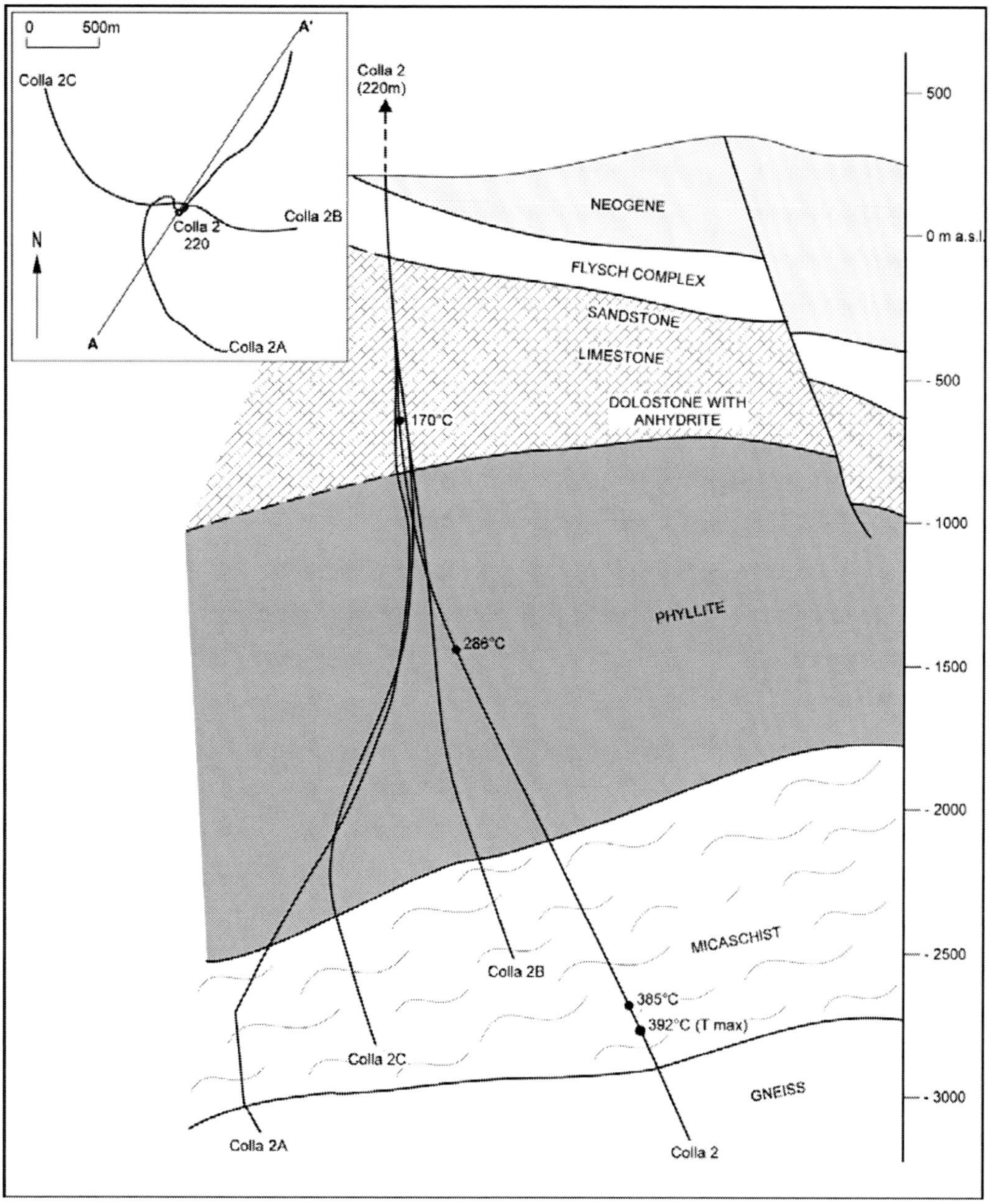

Figure 13. Example of a possible site for heat extraction from a deep-seated reservoir at Larderello.

References

Acocella, V. *Terranova* 2000, 12, 149-155.

Anovitz, L.M.; Labotka, T.C.; Blencoe, J.G.; Horita, J. *Geochim. Cosmochim. Acta* 2004, 68, 3557-3567.

Atkinson Jr., A.B. Master of Science Thesis, Virginia State University, Blacksburg, Virginia, July 16th 2002, 1-126.

Baldi, P.; Bellani, S.; Ceccarelli, S.; Fiordelisi, A.; Rocchi, G.; Squarci, P.; Taffi, L. *Proceedings World Geothermal Congress*, Florence, 18-31 May 1995a, 2, 1287-1291.

Baldi, P.; Bertini, G.; Ceccarelli, A.; Dini, I.; Ridolfi, A.; Rocchi, G. *Proceedings World Geothermal Congress*, Florence, 18-31 May 1995b, 2, 693-696.

Barberi, F.; Innocenti, F.; Ricci, C.A. Rendiconti *Soc. Ital. Mineralogia e Petrologia*, spec. issue 1971, 27, 169-210.

Barelli, A.; Cappetti, G.; Stefani, G. *Proceedings World Geothermal Congress*, Florence, 18-31 May 1995, 2, 1275-1278.

Batini, F.; Fiordelisi, A.; Graziano F.; Toksoz, M.N. *Proceedings World Geothermal Congress, Florence,* 18-31 May 1995, 2, 817-820.

Batini, F.; Bertini, G.; Gianelli,G.; Pandeli, E.; Puxeddu, *M. Mem. Soc. Geol.* It. 1983, 25, 219-235.

Belkin, H.; De Vivo, B.; Gianelli, G.; Lattanzi, *B. Geothermics* 1985, 14, 59-72.

Bertani, R.; Cappetti, G. *Proceedings World Geothermal Congress*, Florence, 18-31 May 1995 ,2, 1735-1740.

Bertini, G.; Cappetti, G.; Dini, I.; Lovari, F. *Proceedings World Geothermal Congress*, Florence, Italy, May 18-31, 1995, 2, 1283-1286.

Bertini, G; Casini, M; Gianelli, G.; Pandeli, *E. Terra Nova*, 2006, 18, 163-169.

Bischoff, J.L.; Pitzer, K.S. *Amer. J. Sci.* 1989, 289, 217-248.

Bischoff, J.L.; Rosenbauer, R.J. *Geochim. Cosmochim. Acta*, 1988, 52, 2121-2126.

Boccaletti, M.; Coli, M.; Eva, C.; Ferrari, G.; Giglia G.; Lazzarotto, A.; Merlanti, F.; Nicolich R.; Papani, G.; Postpischl D. *Tectonophysics* 1985, 117, 7-38.

Boccaletti, M.; Gianelli, G.; Sani, F. *Tectonophysiscs* 1996, 270, 127-143.

Bodnar, *R.J. Pure & Appl. Chem.* 1995, 67, 873-880.

Bodnar, R.J.; Burnham, C.W.; Sterner, S.M. (1985). *Geochim. Cosmochim. Acta*, 1985, 49, 1861-1873.

Bowers, T.S.; Helgeson, H.C. *Geochim. Cosmochim. Acta* 1983, 47, 1247-1275.

Caporali A.; Martin S.; Massironi M. *Geophys. J. Int.* 2003, 155, 254-268.

Calore, C.; Celati, R.; Gianelli, G.; Norton, D.; Squarci, P. *In Energia geotermica: prospettive aperte dalle ricerche del CNR.* Sottoprogetto "Energia Geotermica", Secondo seminario informativo, Roma, 16-19 giugno 1981, 218-225.

Cameli, G.M.; Batini, F.; Dini, I.; Lee, J.M.; Gibson R.L.; Toksoz M.N. *Proceedings World Geothermal Congress*, Florence, 18-31 May 1995, 2, 821-826.

Cappetti, G.; Parisi, L.; Ridolfi, A.; Stefani, G. *Proceedings World Geothermal Congress, Florence, 18-31 May* 1995 2, 1997-2000.

Carter N.C.; *Tsenn M. Tectonophysics* 1987, 136, 27-63.

Cataldi, R.; Lazzarotto, A.; Muffler, P.; Squarci, P.; Stefani, *G. Geothermics*, 1978, 7, 91-131.

Cathelineau, M.; Marignac, Ch. ; Boiron, M.C.; Gianelli, G.; Puxeddu, M. *Geochimica* et *Cosmochimica Acta* 1994, 58, 1083-1099.

Cavarretta, G.; Gianelli, G.; Puxeddu, M. *Econ. Geol.* 1982, 77, 1071-1084.

Cavarretta, G.; Puxeddu, M. *Econ. Geol.* 1990, 85, 1236-1251.

Cavarretta, G.; Puxeddu, M. *N. Jb. Mineral. Abh.* 2001, 177, 77-112.

Cathelineau, M.; Marignac, C.; Boiron, M.C.; Gianelli, G.; Puxeddu, M. *Geochim. Cosmochim. Acta* 1994, 58, 1083-1099.

Ceccarelli, A.; Celati, R.; Grassi, S.; Minissale, A.; Ridolfi, *A. Geothermics* 1987,16,505-515.

Celati, R.; Cappetti, G.; Calore, C.; Grassi, S. *Geothermics* 1991, 20, 119-133.

Chou, I-M. Geochim. *Cosmochim. Acta,* 1987, 51, 1965-1975.

D'Amore, F. In ISRH-89 (Htdrothermal Reactions), *3rd International Simposium* , September 12-15. 1989, Frunze, Kirghizia, U.S.S.R., Abstract p.66.

D'Amore, F.; Bolognesi, L.. *Geothermics* 1994, 23, 21-32.

D'Amore, F.; Truesdell, A. H. In *Proc. 5th Workshop on Geothermal Reservoir Engineering,* Stanford, CA, 1979, 283-297.

Del Moro, A.; Puxeddu, M.; Radicati di Brozolo, F.; Villa, I. *Contrib. Mineral. Petrol.* 1982, 81, 340-349.

Della Vedova, B., Marson, I., Panza, G.F. and Suhadolc, P. *Tectononophysics* 1991, 195, 311-318.

Dini, A.; Gianelli, G.; Puxeddu M.; Ruggieri G. *Lithos* 2005, 81, 1-31.

Elter, P.; Giglia, G.; Tongiorgi, M.; Trevisan, L. *Boll. Geofisica Teorica e Applicata* 1975, 17, 3-18.

Ferrarri, L.; Conticelli, S.; Burlamacchi, L.; Manetti, P. (1996) *Acta Vulcanologia* 1996, 8, 41-56.

Fiordelisi, A.; Mackie, R.L.; Madden, T.; Manzella, A.; Rieven, S.A. *Proceedings World Geothermal Congress*, Florence, 18-31 May 1995, 2, 893-897.

Foley, J.E.; Toksoz, M.N.; Batini, F. *Geophys. Res. Letters* 1992, 19, 5-8.

Fornier, R.O. In Volcanism in Hawaii; Decker, R.W.; Wright, T.L; Stauffer, P.H.; Ed.; U.S. *Geol. Survey Prof. Paper* 1987, 1350(2), 1487-1506.

Fournier, R.O. Geophys. *Res. Lett.*, 1991, 18 94, 955-958.

Fournier, R.O. *Economic Geology* 1999, 94, 1193-1209.

Geiger, S.; Driesner, T.; Heinrich, C.A.; Matthai, S. *Trans. Porous Media* 2006, 63, 399-434.

Gianelli G. Boll. *Soc. Geol.* Ital., 1985, 105, 575-584.

Gianelli, G. In *NEDO's Workshop on Deep-seated and Magma-ambient Geothermal systems,* Tokyo, 1994a, 27-36.

Gianelli G. *Studi Geologici Camerti,* special volume 1994b, 1, 195-200.

Gianelli G. *Mem. Soc. Geol.* It.1994c, 48, 707-713.

Gianelli, G.; Bertini, G. *Boll. Soc. Ital.* 1993, 112, 507-512.

Gianelli G.; Laurenzi, M.A. *Geothermal Resources Council Transactions* 2001, 25, 731-735.

Gianelli, G.; Manzella, A.;Puxeddu, M. *Geothermal Resources Council Transactions* 1996, 20, 287-293.

Gianelli, G.; Manzella, A.; Puxeddu, *M. Tectonophysics* 1997a, 281, 221-239.

Gianelli, G.; Puxeddu, M. *Mem. Soc. Geol.* Ital. 1994, 48, 715-717.

Gianelli, G.; Puxeddu, M.; Batini, F.; Bertini, G.; Dini, I.; Pandeli, E.; Nicolich, R. *Geothermics* 1988, 17, 719-734.

Gianelli, G.; Puxeddu, M.; Ruggieri, G. In: *Water-Rock interaction* 2001, Cidu, R.; Ed., Swets and Zeitlinger, Lisse (ISBN 90 2651 824 2), 2001, 839-842.

Gianelli, G.; Puxeddu, M.; Squarci, P. *Mem. Soc. Geol. Ital.* 1978, 19, 469-476.

Gianelli, G.; Squarci, P.; Capocecera, P.; Baldi, P.; Cappetti, G.; Console, R. *Geothermal Resources Council Transactions* 1997b, 21, 273-275.

Giolito, C. Doctoral (Ph.D.) Thesis, University of Florence, Italy 2005, 1-164.

Hashida, T.; Bignall, G.; Tsuchiya, N.; Takahashi, T.; Tanifuji, K. *Geothermal Resources Transactions* 2001, 25, 225-229.

Ingebritsen, S.E.; Manning, C.E. *Geology* 1999, 27, 1107-1110.

Ingebritsen, S.E.; Manning, C.E. *J. Geochem*. Explor. 2003, 78-79, 1-6.

Kligfield, R.; Hunziker, J.; Dallmeyer, R.D; Schamel, S. *J. Struct. Geol.* 1986, 8, 781-798.

Magro, G.; Ruggieri, G., In: Geoitalia 1999, 2° *Forum Italiano delle Scienze della Terra*, Bellaria, Italy 1999, 1, 249-250.

Magro, G.; Ruggieri,; G.; Gianelli, G.; Belliani, S. *J. Geophys. Res*. 2003, 108 ECV 3, 1-12.

Manning, C.E.; Ingebritsen, S.E. *Rev. Geophys*., 1999, 37, 127-150.

Marini, L.; Manzella A., J. *Volc. Geoth. Res*. 2005, 148, 81-97.

Minissale, A. (1991) The Larderello geothermal field: a review. *Earth Science Reviews* **31**, 133-151.

Minissale, A.; Magro, G.; Vaselli, O.; Verrucchi, C.; Perticone, I. *Journal of Volcanology and Geothermal Research* 1997, 79, 223-251.

Mongelli, F.; Palumbo, F.; Puxeddu, M.; Villa, I.M.; Zito, G.; *Mem. Soc. Geol. Ital.* 1998, 52, 305-318

Montone, P.; Amato, A.; Pondrelli, S. 1999, *J. Geophys. Res*., **104** B11, 25595-25610.

Morrow, C.A.; Moore, D.E.; Lockner, D.A. *Journal of Geophysical research* 2001, 106, B12, 30551-30560.

Pandeli, E.; Bertini, G.; Castellucci, P. *Boll. Soc. Geol. Ital.* 1991, 110, 621-629.

Pandeli, E.,; Gianelli, G.; Puxeddu, M.; Elter, F.M. *Mem.Soc. Geol. Ital.* 1994, 48, 627-654.

Panichi, C.; Celati, R.; Noto, P.; Squarci, P.; Taffi, L.; Tongiorgi, E. In *Isotope Techniques in Groundwater Hydrology*, 1974, 2, 4-28.

Panza, G.F.; Suhadolc, P. *Tectonophysics* 1990, 182, 39-46.

Petrucci, E.; Gianelli, G.; Puxeddu, M.; Iacumin, P. *Geothermics* 1994, 23, 327-337.

Petrucci, E.; Sheppard, S.M.F.; Turi, B. J. *Volc. Geotherm. Res*. 1993, 59, 145-160

Petracco, C.; Squarci, P. In: Proceedings 2nd U.N. *Symposium on Development and Use of Geothermal Resources*, San Francisco U.S.A. 1976, 1, 521-530.

Pichavant, M.; Kontak, D.J.; Valencia Herrera, J.; Clark, A.H. *Contrib. Mineral. Petrol.* 1988, 100, 300-324.

Poli, G.; Manetti, P.; Tommasini, S. *Periodico Mineralogia*, 1989, 58, 109-126.

Pruess, K.; Celati, R.; Calore, C.; Cappetti, G. In *Proc. 12th Workshop on Geothermal Reservoir Engineering*, Stanford, CA, 1987, 89-96.

Reutter, K.J.; Giese, P.; Closs, H. *Tectonophysics* 1980, 64, 1-9.

Rowland, J.V. ; Sibson, R.H. *Geofluids* 2004, 4, 259-283.

Ruggieri, G.; Cathelineau, M.; Boiron; M.C.; Marignac, C. *Chemical Geology* 1999, 154, 237-256.

Ruggieri, G.; Gianelli, G. *Proceedings World Geothermal Congress, Florence*, 18-31 May 1995, 2, 90-94.

Ruggieri, G.; Gianelli, G. *Geothermics* 2002, 31, 443-474.

Ruggieri G.; Giolito C.; Gianelli G.; Manzella A.; Boiron M.C. *Geothermics*, 2004, 33, 675-692.

Serri G.; Innocenti, F.; Manetti, P. *Tectonophysics* 1993, 223,117-147.

Sibson, R.H. *Tectonophysics* 1992, 18, 1031-1042.

Takahashi, T.; Tanifuji, K.; Stafford C.; Hashida, T. *JSME international Journal*, series A 2002, 46, 24-29.

Tenthorey, E.; Fitz Gerard, J.D. *Earth and planetary Science Letters* 2006, 247, 117-129.

Truesdell, A.H.; Halizip, J.R.; Armannson, H.; D'Amore, F *Geothermics*, 1989, 18, 295-304.

Truesdell, A. H. In *Proc. 16th Workshop on Geothermal Reservoir Engineering Stanford University,* Stanford,California, January 23-25 1991, SGP-TR-134, 15-20.

Tsuchiya, N.; Hirano, N.; Bignall, G.; Nakatsuka, K. In Water-Rock Interaction 2001, Cidu, R.; Ed.; Swets and Zeitlinger, Lisse (ISBN 90 2651 824 2), 209-212..

Turi, B.; Taylor, H.P. *Contrib. Mineral. Petrol.* 1976 **55**, 1-31.

Van Bergen, M.J.; Ghezzo, C.; Ricci, C.A. J. Volc. *Geotherm. Res.* 1983, 19, 1-35.

Villa, I.; Puxeddu, M. *Contrib. Mineral. Petrol.* 1994, 15, 415-426.

White, D. E. (1981) Active geothermal systems and hydrothermal ore deposits. *Econ. Geol. 75th Anniversary Volume*, **392**-423.

White, D. E.; Muffler, L. J. P.; Truesdell, A. H *Econ. Geol.* 1971, 66, 75-97.

In: Geothermal Energy Research Trends
Editor: Herman I. Ueckermann, pp. 87-107

ISBN 1-60021-683-8

Chapter 4

INSTABILITY AND BIFURCATION OF HORIZONTALLY HOMOGENEOUS FLOWS IN GEOTHERMAL SYSTEMS

A. T. Il'ichev* * and G. G. Tsypkin
Steklov Mathematical Institute, Gubkina Str. 8 119991
Moscow, Russia
Institute for Problems of Mechanics, Vernadskogo pr. 101 119526
Moscow, Russia

Abstract

High-temperature geothermal reservoir is under consideration, consisting of two high-permeability layers, which are separated by a low-permeability stratum. The thermodynamic conditions are assumed to imply that the upper and lower high-permeability layers are filled in by water or by vapor, respectively. The stable stationary regimes of vertical phase flow between water and vapor layers in the low-permeability stratum may exist. We give possible types of transition to instability of the vertical flows in such a system under the condition of smallness of the advective heat transfer in comparison with the conductive one. It is found that in the generic case there exist three different scenarios of the instability onset of the stationary vertical flows. They are accompanied by the bifurcations of solutions describing the destabilizing vertical flows. The possible scenarios of the evolution of the system over the threshold of instability are discussed.

1. Introduction

Geothermal reservoirs are related with thermal anomalies in lithosphere. These anomalies can be caused by inhomogeneities in a structure of lithosphere itself as well as have a hydrodynamic nature. In the latter case steam-water countercurrent flow steady-state regimes

*E-mail address: ilichev@mi.ras.ru.

arise in permeable rock driven by the application of a heat flux and gravity, the so called heat pipe regimes [1, 6, 20, 21]. The heat pipe provides enhancement of the heat transfer over that of conduction in liquid-saturated or vapor-saturated hot rock due to the convective contributions of the both phases. Natural observations and experimental investigations show that the heat pipe is the most effective mechanism of the heat transport in geothermal reservoir in the vertical direction. A mathematical description of the heat transport in the heat pipe was developed in [1, 20, 22]. The possibilities of energy extraction from a geothermal reservoir considerably depends on heat pipe properties. In this context the importance of heat pipes in the functioning of geothermal sources of energy is hard to overestimate.

The formation of the heat pipe or vapor-liquid counterflow zone evidently represents the result of the evolution of gravitationally unstable regime, when the heavier fluid moves downwards and the lighter fluid moves upwards. Meanwhile the existence of water over steam in geothermal systems is well known in natural systems [8, 17, 21]. In the classical paper [16] it was shown theoretically that such a configuration in a low permeability rock (permeability is less than 10^{-17} m^2) can be stable, when the base (unperturbed) state describes the motionless equilibrium of the water layer over the vapor one. The model under consideration in this paper assumes the existence of a phase transition interface between water and vapor layers. The result of this work was extended to the case when there is a motion through the interface in [19]. In this setting the base state described either the vapor motion upwards with its following condensation at the interface, or water motion downwards with its following vaporization at the interface. This regime itself still has nothing to do with the heat pipe since it is characterized by a significant temperature gradient while exists stably for some range of physical parameters of the geothermal reservoir. Yet the loss of stability leads to destruction of the sharp interface and the possible origin of convective regimes which can be treated as a heat pipe. However it is not clear *apriori* what kind of regime arises when the interface is destructed as a result of the loss of stability. The question about secondary regimes in theory of hydrodynamic stability is of fundamental importance and complex enough to be studied by analytical methods. In this connection the investigation of the type of transition is important and may provide the information about possible secondary regimes [4].

In the framework of classical fluid dynamics the equilibrium of heavy fluid over the lighter one is always unstable. The instability in question is known as the Rayleigh-Taylor instability (see e. g. [2]). Yet, when supplementary physical mechanisms have to be taken into consideration in a hydrodynamic model, such an equilibrium is possible. The geothermal reservoirs represent the example of physical system where similar mechanisms have a stabilizing effect [9, 16, 19]. The existence of such a configuration is explained from the thermodynamic point of view by a considerable temperature gradient typically taking place in geothermal system. In this case the thermodynamic conditions at large depths correspond to vapor, and at smaller depths – to water.

In [16] the example of the geothermal reservoir of the homogeneous vertical extend embedded in the rigid rock is treated. In such a reservoir the base (unperturbed) state describes the motionless equilibrium of the water layer over the vapor one. The stability of such a configuration for some range of values of the permeability of the geothermal reservoir was shown by means of numerical analysis of the dispersion relation.

In [19] the more general case of a geothermal system is analyzed in a sense that the phase motion and phase transition in a base state are permitted. A high-temperature geothermal reservoir was considered, consisting of two high-permeability layers, which are separated by a low-permeability rock. It was assumed that the thermodynamic conditions imply that the upper and lower high-permeability layers are filled in by water and by vapor, respectively. Then in the low-permeability layer there exists the phase transition interface, separating domains occupied by water and vapor. The geometry of the reservoir is presented in Fig.1. In dependence on values of pressure in the high-permeability layers, either the regime of vaporization, when the water moves downwards, or the regime of condensation, corresponding to the vapor motion upwards, take place. In the rest state phase transitions are absent, and the pressure distribution in the high-permeability layers coincides with the hydrostatic one.

Calculations showed that there exist the regimes of motion with phase transitions, when the water layer over the vapor layer in geothermal system occurs to be stable also for permeability values which exceed the maximal value got in [16] by the order of magnitude. The new threshold value of the permeability (10^{-16} m^2) explains the fact of stability of a lot of geothermal systems, observed in nature.

The character of the arising secondary flows depends on the type of instability and, therefore, it seems to be important to investigate in detail the possible types of transition to instability. In this paper we present the results of the analytic investigation of stability of a flow in the model of the geothermal system treated in [9, 19]. The criteria of the transition to instability satisfying the principle of exchange of stabilities (when the time exponent of the least stable normal mode is zero at the margin of stability) are obtained in explicit form. It is established that the transition to instability of the interface under the variation of physical parameters takes place by means of one of the following four mechanisms, determined by the wave numbers of the most unstable modes at the threshold of instability:

(i) comes spontaneously for all wave numbers of perturbations (degenerate case);
(ii) the most unstable mode has an infinitely large wave number;
(iii) the most unstable mode is infinitely long;
(iv) the most unstable mode is of a finite length.

The loss of stability of the horizontal phase transition interface is accompanied by bifurcations leading to the branching of base regimes, describing horizontally homogeneous vertical phase flows. The particular feature of the transition to instability of vertical flows in the cases (ii-iv) consists in disappearance of the regimes, losing their stability and origin of new unstable vertical flows with plane phase transition interface.

Each type of instabilities enumerated above is characterized by a certain weakly nonlinear stage of a development of instability with the corresponding governing nonlinear equation for perturbations. The governing equation for the case (ii) of the first destabilized infinitely short waves is still unknown for us. The equation describing the band of weakly unstable modes in the case (iii) occurs to be the well known KPP equation [10, 11]. As far as we know this equation was not treated before as modulation equation for weakly nonlinear stage of development of instability and we present its derivation in Appendix. The governing equation for weakly unstable band of modes in the case (iv) is evidently the Ginzburg-Landau type equation which is extensively studied in literature as a modulation equation for perturbations (see e.g. [3, 5, 13])

The paper organized as follows. In sec. 2 we give formulation, in sec. 3 we describe base regimes, in sec. 4 the dispersion relation is derived. In sec. 5 the types of transition to instability in the degenerate case (i) (when the phase transition interface of the base regime is located exactly in the middle of the low-permeability layer) and in the cases (ii-iv) are described. In section 5 we discuss the results and give our conclusions. In Appendix we present the derivation of modulation equation for the case (iii) in general case.

2. Formulation of the Problem

Notations

Latin symbols

a	$[\mathrm{m^2\,s^{-1}}]$	thermal diffusivity
C	$[\mathrm{J\,K^{-1}\,kg^{-1}}]$	specific heat at constant pressure
g	$[\mathrm{m\,s^{-2}}]$	acceleration of gravity
h	[m]	location parameter of the interface
k	$[\mathrm{m^2}]$	permeability
L	[m]	thickness of the low permeable stratum
m	[1]	porosity
P	[Pa]	pressure
q	$[\mathrm{J\,kg^{-1}}]$	specific heat of phase transition
t	[s]	time
T	[K]	temperature
V	$[\mathrm{m\,s^{-1}}]$	speed of the phase transition interface
x	[m]	vertical coordinate

Greek symbols

η	[m]	perturbation of the interface
κ	$[\mathrm{m^{-1}}]$	wave number
λ	$[\mathrm{W\,m^{-1}\,K}]$	thermal conductivity
μ	[Pa s]	viscosity
ρ	$[\mathrm{kg\,m^{-3}}]$	density
σ	$[\mathrm{s^{-1}}]$	spectral parameter

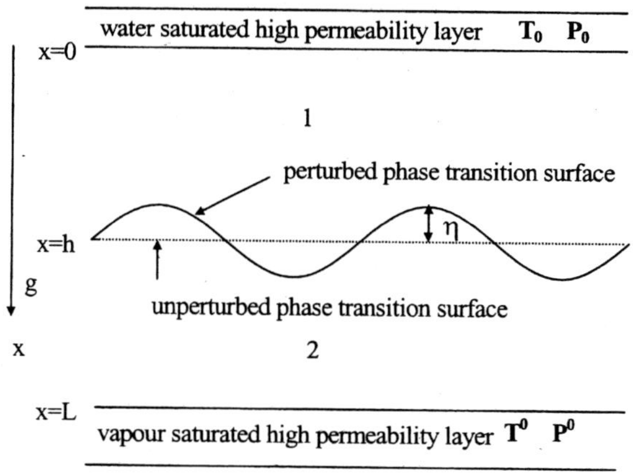

Figure 1. Geometry of the problem; 1 – water domain in the low-permeability stratum, 2 – vapour domain in the low-permeability stratum.

Subscripts

1	water domain
2	vapor domain
n	normal
s	skeleton of porous medium
v	vapor
w	water
0	boundary value at $x = 0$
$+$	right behind of the interface in the vapor saturated domain
$-$	right ahead the interface in the water domain
$*$	at the phase transition front

Superscript

0	boundary value at $x = L$

We consider high-temperature geothermal reservoir, consisting of two high-permeability horizontal layers, separated by a low-permeability rock having sufficiently large horizontal extension. The low-permeability rock is covered by a strip $-\infty < y < \infty$, $0 < x < L$ with the vertical x-axis directed downwards. The upper and lower high-permeability layers are occupied by water and vapor, respectively. Inside the low-permeability stratum there exists the phase transition interface, separating domains occupied by water and vapor, respectively (Fig.1).

Processes of heat and mass transfer in the framework of equilibrium thermodynamics can be described by mass and energy conservation laws, the Darcy law for water and vapor with allowance for gravity, the equations of state and thermodynamic relations [15]. The water and vapor are assumed to be incompressible. Then the basic system of equations for the two domains of the low-permeability rock separated by the interface read

$$
\begin{aligned}
&\text{div}\, v_j = 0, \qquad v_j = -\frac{k}{\mu_j}(\,\text{grad}\, P - \rho_j g \mathbf{e}_x),\\
&(\rho C)_{1,2}\frac{\partial T}{\partial t} + \rho_j C_j v_j\, \text{grad}\, T = \text{div}\,(\lambda_{1,2}\,\text{grad}\, T),\\
&\lambda_{1,2} = m\lambda_j + (1-m)\lambda_s,\\
&(\rho C)_{1,2} = m\rho_j C_j + (1-m)\rho_s C_s, \quad j = v, w.
\end{aligned}
\tag{2.1}
$$

The conservation of mass and energy across the interface are formulated as the conditions of the thermodynamical equilibrium jump of the water saturation function [7, 18]. These relations have the form

$$
\begin{aligned}
&m(1-\frac{\rho_v}{\rho_w})V_n = \frac{k}{\mu_v}\frac{\rho_v}{\rho_w}(\,\text{grad}\, P)_{n+} - \frac{k}{\mu_w}(\,\text{grad}\, P)_{n-} +\\
&\qquad\qquad + \frac{k}{\mu_w}\rho_w g(1 - \frac{\mu_w}{\mu_v}\frac{\rho_v^2}{\rho_w^2}),\\
&mq\rho_w V_n = \lambda_-(\,\text{grad}\, T)_{n-} - \lambda_+(\,\text{grad}\, T)_{n+} - \frac{kq\rho_w}{\mu_w}((\,\text{grad}\, P)_{n-} - \rho_w g),\\
&T_+ = T_- = T_*, \qquad P_+ = P_- = P_*, \quad \ln\frac{P_*}{P_a} = A + \frac{B}{T_*},\\
&A = 12.512, \quad B = -4611.73\text{K}, \quad P_a = 10^5\text{Pa}.
\end{aligned}
\tag{2.2}
$$

The second term in the left-hand side of the heat transfer equation in (2.1) describes the heat advective transfer while the right-hand side describes the heat conductive transfer.

Boundary conditions for the pressure and temperature in high-permeability layers are given by

$$
x = 0:\ P = P_0, \quad T = T_0; \qquad x = L:\ P = P^0, \quad T = T^0. \tag{2.3}
$$

Our principal restriction is that we assume that the conductive heat transfer in the system considerably exceeds the advective one. This restriction, nevertheless, permits to treat a lot of natural geothermal systems. This statement is based on the following estimates. For the water domain, combining Darcy's law with the heat conservation equation, we obtain the dimensionless parameter that specifies the ratio of advective and conductive terms [19]:

$$
\frac{\rho_w C_w}{\mu_w \lambda_1}k(\delta P - \rho_w g l), \tag{2.4}
$$

where l is the characteristic length scale and δP is dynamical pressure variation. For other notations see table at the beginning of sec. 2 of the paper. Permeability and pressure may vary strongly, at the same time the other physical parameters vary slightly. Therefore, after substitution of the characteristic values of parameters in (2.4) the condition of smallness of advective transfer in the water domain can be written in the following form

$$
k \mid \delta P - \rho_w g l \mid \ll 10^{-10}\text{N}, \tag{2.5}
$$

Analogously, one has for the vapor domain

$$k \mid \delta P - \rho_v g l \mid \ll 3 \cdot 10^{-9}\mathrm{N}. \tag{2.6}$$

The criteria (2.5), (2.6) occur to be valid for a lot of flows in natural geothermal systems.

3. Base Flows

Under the condition of smallness of the advective energy transfer the system of equations for the unknown pressure and temperature in the water and vapor domains are given by

$$P''(x) = 0, \qquad T''(x) = 0.$$

Conservation laws of mass and energy on the interface for the case under consideration read

$$\frac{\mu_w}{\mu_v}\frac{\rho_v}{\rho_w} P'_+ - P'_- + \rho_w g(1 - \frac{\mu_w}{\mu_v}\frac{\rho_v^2}{\rho_w^2}) = 0,$$

$$\lambda_+ T'_+ - \lambda_- T'_- + \frac{kq\rho_w}{\mu_w}(P'_- - \rho_w g) = 0,$$

where prime denotes differentiation with respect to x. These relations, along with the conditions of thermodynamic equilibrium, constitute the full system of boundary conditions across the interface.

The solutions in the water and vapor domains have the form

$$P = P_0 + \frac{P_* - P_0}{h}x, \qquad T = T_0 + \frac{T_* - T_0}{h}x$$
$$P = \frac{P^0 - P_*}{L-h}x + \frac{LP_* - hP^0}{L-h}, \quad T = \frac{T^0 - T_*}{L-h}x + \frac{LT_* - hT^0}{L-h}, \tag{3.1}$$

respectively. Substituting (3.1) into the stationary relations on the interface we get the system of equations for the unknowns T_*, P_* and h:

$$\frac{\mu_w}{\mu_v}\frac{\rho_v}{\rho_w}\frac{P^0 - P_*}{L-h} - \frac{P_* - P_0}{h} + \rho_w g\left[1 - \frac{\mu_w}{\mu_v}\frac{\rho_v^2}{\rho_w^2}\right] = 0,$$
$$\lambda_+ \frac{T^0 - T_*}{L-h} - \lambda_- \frac{T_* - T_0}{h} + \frac{kq\rho_w}{\mu_w}\left[\frac{P_* - P_0}{h} - \rho_w g\right] = 0,$$
$$P_* = f(T_*) = P_a \exp\left(A + \frac{B}{T_*}\right). \tag{3.2}$$

From the first equation the vaporization pressure on the interface can be expressed as a function of dimensionless location parameter $H = h/L$:

$$\frac{P_*}{P_0} = \frac{(P^0/P_0)(c_1/(1-H)) + 1/H + c_2}{c_1/(1-H) + 1/H} \equiv F_1(H),$$

$$c_1 = \frac{\mu_w}{\mu_v}\frac{\rho_v}{\rho_w}, \qquad c_2 = \frac{\rho_w g L}{P_0}\left[1 - \frac{\mu_w}{\mu_v}\frac{\rho_v^2}{\rho_w^2}\right].$$

The temperature is expressed from the third equation as a function of the pressure:

$$\frac{T_*}{T_0} = \frac{B/T_0}{\ln P_*/P_0 - \ln P_a/P_0 - A} = \frac{B/T_0}{\ln F_1(H) - \ln P_a/P_0 - A} \equiv F_2(H).$$

This makes it possible to reduce the system to the unique equation for the unknown parameter $H = h/L$

$$G(H) \equiv \left[\frac{T^0}{T_0} - F_2(H)\right]\frac{1}{1-H} - [F_2(H) - 1]\frac{1}{H} + \\ + \frac{kq\rho_w P_0}{T_0\lambda}\left[\frac{F_1(H) - 1}{H} - \frac{\rho_w g L}{P_0}\right] = 0. \qquad (3.3)$$

4. Dispersion Relation

For investigation of normal stability of a stationary flow with the phase transition interface we linearize the basic equations. The perturbations of the pressure and temperature satisfy

$$\Delta P = 0, \qquad \frac{\partial T}{\partial t} = a_{1,2}\Delta T, \qquad 0 < x < h, \quad h < x < L,$$
$$a_{1,2} = \frac{\lambda_{1,2}}{(\rho C)_{1,2}}, \qquad \Delta = \frac{\partial^2}{\partial x^2} + \frac{\partial^2}{\partial y^2}. \qquad (4.1)$$

For the sake of simplicity we assume that the specific heat and heat conductivity depend only on the corresponding parameters of the skeleton: $a = a_1 = a_2$.

The boundary conditions for perturbations read

$$P = 0, \ T = 0, \qquad \text{at } x = 0, \ L,$$
$$P_- = P_+ + \frac{P_0}{L}\Gamma_1\eta, \ \ \Gamma_1 = \frac{1}{H} + \frac{P_1 H - P_f}{H(1-H)}, \quad \text{at } x = h,$$
$$T_- = T_+ + \frac{T_0}{L}\Gamma_2\eta, \ \ \Gamma_2 = \frac{1}{H} + \frac{T_1 H - T_f}{H(1-H)}, \quad \text{at } x = h,$$
$$P_- = \left(\frac{\partial f(T)}{\partial T}\right)_{T=T_*} T_- + \left[\left(\frac{\partial f(T)}{\partial T}\right)_{T=T_*}\left(\frac{\partial T}{\partial x}\right)_- - \left(\frac{\partial P}{\partial x}\right)_-\right]\eta =$$
$$= -\frac{P_0 B}{T_0^2}\Gamma T_- - \frac{P_0}{L}\Gamma_0\eta, \ \ \Gamma = \frac{P_f}{T_f^2}, \ \ \Gamma_0 = \frac{B}{T_0}\Gamma\frac{T_f - 1}{H} + \frac{P_f - 1}{H}, \ \ \text{at } x = h, \qquad (4.2)$$
$$m(1-R)\frac{\partial \eta}{\partial t} = \frac{k}{\mu_v}R\left(\frac{\partial P}{\partial x}\right)_+ - \frac{k}{\mu_w}\left(\frac{\partial P}{\partial x}\right)_-, \quad R = \frac{\rho_v}{\rho_w}, \quad \text{at } x = h,$$
$$mq\rho_w\frac{\partial \eta}{\partial t} = \lambda_-\left(\frac{\partial T}{\partial x}\right)_- - \lambda_+\left(\frac{\partial T}{\partial x}\right)_+ - \frac{kq\rho_w}{\mu_w}\left(\frac{\partial P}{\partial x}\right)_-, \quad \text{at } x = h,$$

where $x = h + \eta(t, y)$ is the interface equation, $T_f = T_*/T_0$, $T_1 = T^0/T_0$, $P_f = P_*/P_0$, $P_1 = P^0/P_0$.

Assuming, that the unknown functions can be represented as

$$\{P(x,y,t), T(x,y,t), \eta(y,t)\} = \{\hat{P}(x), \hat{T}(x), \hat{\eta}\} \exp(\hat{\sigma}t + i\hat{\kappa}y),$$

we get the dispersion equation from the condition of existence of the non-trivial solution:

$$\begin{aligned} F(\sigma,\kappa) \equiv & \alpha \coth[\alpha(1-H)]\Big[\frac{\omega_w}{a}\Gamma_3\kappa \coth(\kappa H) - (1-R)\sigma + \\ & + \frac{\omega_v}{a}\Gamma_4 R\kappa \coth[\kappa(1-H)]\Big] + \alpha \coth(\alpha H)\Big[\frac{\omega_v}{a}\Gamma_5 R\kappa \coth[\kappa(1-H)] + \\ & + \frac{\omega_w}{a}\Gamma_0\kappa \coth(\kappa H) - (1-R)\sigma\Big] + \frac{\omega_v B}{T_0}\frac{mq\rho_w R}{\lambda T_0}\Gamma\kappa\Big[\sigma \coth[\kappa(1-H)] + \\ & + \sigma\frac{\mu_v}{\mu_w}\coth(\kappa H) + \Gamma_1\kappa\frac{\omega_w}{a}\coth(\kappa H)\coth[\kappa(1-H)]\Big] = 0. \end{aligned} \tag{4.3}$$

Here

$$\Gamma_3 = \frac{B}{T_0}\Gamma\frac{T_1 - T_f}{1-H} + \frac{P_f - 1}{H}, \qquad \Gamma_4 = \frac{B}{T_0}\Gamma\frac{T_1 - T_f}{1-H} + \frac{P_1 - P_f}{1-H},$$

$$\Gamma_5 = \frac{B}{T_0}\Gamma\frac{T_f - 1}{H} + \frac{P_1 - P_f}{1-H}, \qquad \alpha = \sqrt{\kappa^2 + \sigma}, \quad \kappa = \hat{\kappa}/L,$$

$$\sigma = a\hat{\sigma}/L^2, \quad \omega_w = \frac{P_0 k}{m\mu_w}, \quad \omega_v = \frac{P_0 k}{m\mu_v}.$$

From (4.3) one gets the asymptotic $\sigma = \sigma_0|\kappa|$, $\kappa \to \pm\infty$,

$$\begin{aligned} \sigma_0 \ = \ & \left[\frac{\omega_v}{a}R(\Gamma_4 + \Gamma_5) + \frac{\omega_w}{a}(\Gamma_0 + \Gamma_3) + \frac{\omega_w\omega_v}{a}\frac{B}{T_0}\frac{mq\rho_w R}{\lambda T_0}\Gamma\Gamma_1\right] \times \\ & \left[2(1-R) - \frac{\omega_v}{a}\frac{B}{T_0}\frac{mq\rho_w R}{\lambda T_0}\left(1 + \frac{\mu_v}{\mu_w}\right)\Gamma\right]^{-1}. \end{aligned} \tag{4.4}$$

5. Transition to Instability

In this section the possible types of the exchange of stabilities of the system are considered. In the degenerate case of the location of phase transition interface $x = h = L/2$ the onset of instability takes place simultaneously for all normal modes. In the general case $h \neq L/2$ there are three different mechanisms of the exchange of stabilities.

Putting $\sigma = 0$ in the dispersion equation (4.3) and dividing it by the non-vanishing function $\kappa^2 \coth^2(1-H)\kappa$, one gets the equation

$$\Gamma_0\omega_w Z^2(\kappa) + \left(\omega_w\Gamma_3 + \omega_v\Gamma_5 R + \frac{\omega_v\omega_w B mq\rho_w R}{\lambda T_0^2}\Gamma\Gamma_1\right) Z(\kappa) + \omega_v\Gamma_4 R = 0 \tag{5.1}$$

where

$$Z(\kappa) = \frac{\coth H\kappa}{\coth(1-H)\kappa}.$$

The quadratic equation (5.1) has a solution

$$Z(\kappa) = Z_{1,2} = d \pm \sqrt{d^2 - b}, \tag{5.2}$$

where
$$d = -\frac{1}{2\Gamma_0}\left(\Gamma_3 + \frac{\omega_v}{\omega_w}\Gamma_5 R + \frac{\omega_v B m q \rho_w R}{\lambda T_0^2}\Gamma\Gamma_1\right), \quad b = \frac{\omega_v}{\omega_w}\frac{\Gamma_4}{\Gamma_0}R.$$

In the left hand side of the equality (5.2) stands the even and monotonous for $\kappa > 0$ function $Z(\kappa)$, $\min[(1-H)/H, 1] \leq Z(\kappa) \leq \max[(1-H)/H, 1]$. The values of $Z(\kappa)$ satisfying (5.2) are real if the expression under the square root sign is positive. Instability takes place if the equation (5.2) has finite simple real roots (their number is even because of the symmetry of the function $Z(\kappa)$), and the threshold of instability is attained when the real roots of this equation arise. In the absence of the real roots of the equation (5.2), their origin under variation of the physical parameters may occur in the following cases:

- $D = d^2 - b > 0$, $d < \min[(1-H)/H, 1]$ and $Z_2 < Z_1 = \min[(1-H)/H, 1]$;
- $D = d^2 - b > 0$, $d > \max[(1-H)/H, 1]$ and $Z_1 > Z_2 = \max[(1-H)/H, 1]$;
- $D = 0$, $\min[(1-H)H, 1] < d < \max[(1-H)H, 1]$.

5.1. Spontaneous Transition to Instability in the Degenerate Case

First, let us construct the example of the regime when the transition interface is located exactly in the middle of low-permeability layer. Typically, such a regime corresponds to the equilibrium base state without phase motion (filtration speed equals zero). Let, for example, the pressure in the upper water saturated layer be $P_0 = 10^6$ Pa. Then, one can uniquely find the pressure on the interface. The difference between this pressure and the pressure at any boundary equals the hydrostatic pressure. For $L = 40$ m, for example, $P_* = 1.17418 \cdot 10^6$ Pa. Adding the hydrostatic pressure of the vapor in the vapor saturated domain to this value, we obtain the pressure at the lower boundary at $x = L$: $P^0 = 1.17512 \cdot 10^6$ Pa. These values of pressure correspond to the dynamical rest state. In order to satisfy the thermodynamic conditions it is necessary to choose the temperature distribution with boundary values lying, respectively, in the water and vapor domains in the Clapeyron plane. Consequently, the temperature at the upper boundary at $x = 0$ must be smaller than the temperature of water boiling: $T_0 < T_f(P_0) = 451.7134$ K, and at the lower boundary at $x = L$ the temperature has to be larger than the vaporization temperature: $T^0 > T_f(P^0) = 458.97$ K.

The value of the vaporization temperature T_* on the interface has to belong to the Clapeyron curve at $P_* = 1.17418 \cdot 10^6$ Pa and, therefore, this temperature can be uniquely determined and it equals $T_* = 458.93$ K. Arbitrariness exists only in a choice of the temperature gradient. For example, putting $T_0 = 450$ K and using the absence of phase transitions (continuity of derivatives of temperature across the interface) and also the known value of the temperature at the interface, we get the following value of the temperature at the lower boundary $x = L$: $T^0 = 467.86$ K. Let us note, that this solution is realized at all values of the permeability coefficient.

The dynamic equilibrium or rest state solution takes place at the prescribed boundary values of the pressure and when the temperature coincides with the phase transition temperature on the interface. Variation of the pressure or temperature distribution causes the onset

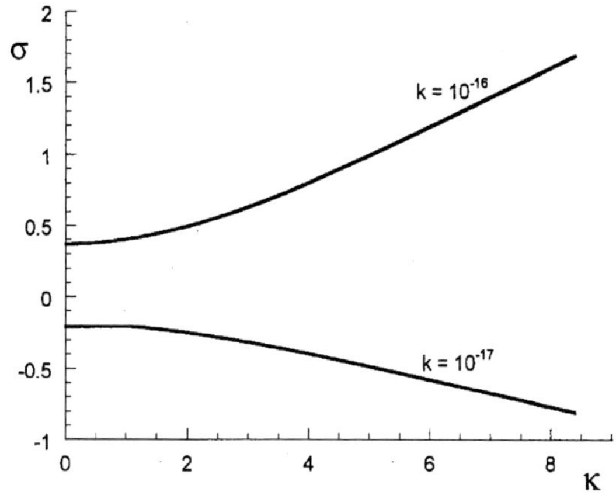

Figure 2. The spontaneous transition to instability of the base regime with $h = L/2$.

of phase transition and phase motion for the stationary solution. It means that the rest state solutions represent isolated solutions in the sense that a small variation of the boundary values leads to a solution with phase motion.

Now, consider the transition to instability of the regime characterized by the phase transition interface location at the point $x = h = L/2$ on the vertical. The set of solutions of (2.1)-(2.3), containing such regimes is not empty as it is demonstrated above. Assume that for some values of the physical parameters the regime in question is stable. Evidently, the necessary condition of stability ($\sigma < 0$ for all κ) occurs to be the negativity of the value σ_0 in (4.4). This is equivalent to the negativity of the numerator

$$\sigma_1 = \omega_v R(\Gamma_4 + \Gamma_5) + \omega_w(\Gamma_0 + \Gamma_3) + \omega_w \omega_v \frac{B}{T_0} \frac{mq\rho_w R}{\lambda T_0} \Gamma \Gamma_1 \tag{5.3}$$

of σ_0, because the denominator of this quantity is always positive. The system comes to the margin of stability when under variation of the parameters the value $\kappa_0 > 0$ arises, such that $\sigma(\kappa_0) = 0$. Substituting $\sigma = 0$ in the dispersion relation (4.3) and taking into account the condition $h = L/2$ ($H = 1/2$) one gets the equality $\sigma_1 = 0$ at all κ. This implies that for $\sigma_1 > 0$ the quantity σ (if real) occurs to be positive exactly for all κ and completely (for all κ) unstable regime is bounded away from the completely stable one by the hypersurface $\sigma_1 = 0$ in the space of parameters. Thus, the equality of σ_1 to zero represents a criterion of the exchange of stabilities for the degenerate stationary regimes, characterized by a certain location of the interface $x = h = L/2$. The instability in this case takes place spontaneously in such a way, that all wavenumbers become unstable at once, when the system crosses the threshold of instability. The example a transition to instability for the regime, described at the beginning of this subsection as the permeability varies is illustrated in Fig. 2. In this example, the instability takes place when $k > k_0$, $k_0 \approx 8.9\,10^{-17}$ m^2 at fixed values of the pressure and temperature on the boundaries of the low-permeability stratum. According to the criterion derived, the quantity σ_1 changes its sign when k begins to exceed k_0.

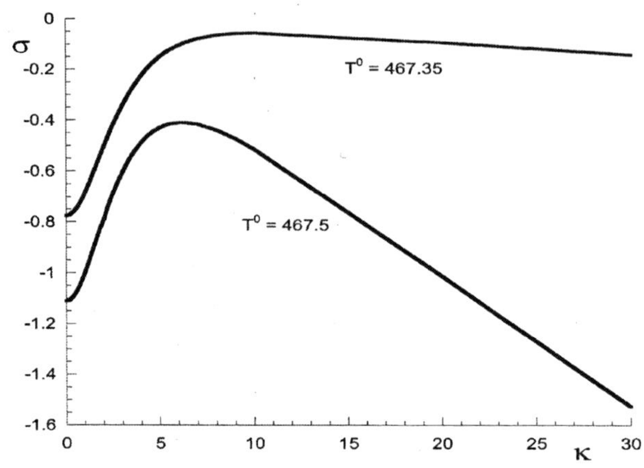

Figure 3. The dispersion curves $\sigma = \sigma(\kappa)$, corresponding to the regimes with $T_0 = 450$ K, $P_0 = 10^6$ Pa, $P^0 = 1.19 \cdot 10^6$ Pa; $L = 40$ m, $k = 10^{-16}$ m^2 ($H < 1/2$). Corresponds to Fig. 6a.

5.2. Origin of Unstable Wave Numbers at Infinity

Let $H < 1/2$, i.e. $\min[(1-H)/H, 1] = 1$. In this case the transition to instability for $d < 1$ takes place at $Z_1 = 1$. The latter equality is attained at $\kappa = \pm\infty$. Therefore, the infinite wavenumber corresponds to the onset of instability in this case. When further variation of parameters leads to the development of instability, the interval of the unstable wave numbers, affiliated with growing normal modes, widens from infinity.

In this setting, as follows from (5.2), the condition of the onset of instability $Z_1 = 1$ is equivalent to the condition $\sigma_1 = 0$, where σ_1 is given by (5.3). The domain of the parameters given by $\sigma_1 < 0$ corresponds to a stable regime and the domain given by $\sigma_1 > 0$ corresponds to an unstable regime. If $H > 1/2$ ($\max[(1-H)/H, 1] = 1$), the onset of instability for $d > 1$ takes place analogously but at $Z_2 = 1$, and the ends of the interval of wave numbers, affiliated with decaying normal modes, come from infinity. The dispersion curves $\sigma = \sigma(\kappa)$ of the most unstable modes of the regimes where the prescribed type of transition to instability is realized are pictured in Fig. 3. The transition to instability in this case is illustrated in Fig. 6a,c.

Numerical experiments show that the type of the transition under consideration is accompanied by decrease either of reservoir temperature or reservoir pressure. When the reservoir temperature (pressure) is large enough the unique stable base solution relating to the flow with plane interface exists. The decrease of temperature (pressure) causes the appearance of a couple of complementary unstable regimes with plane interface. The further decrease of the reservoir temperature (pressure) is accompanied by merging of the stable and unstable solutions which cease to exist. Finally, only one unstable solution, corresponding to the flow with plane interface occurs.

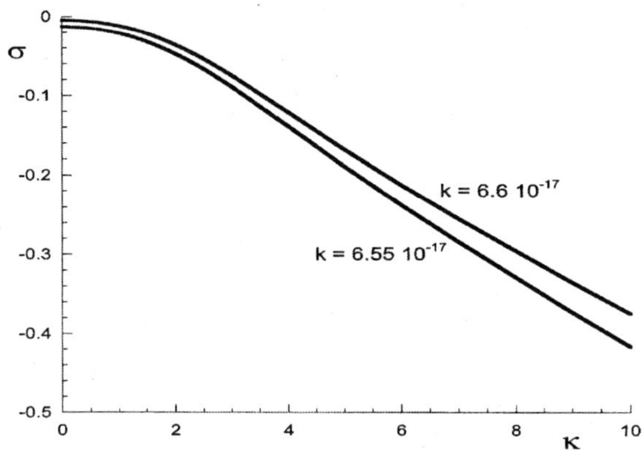

Figure 4. Dispersion curves $\sigma = \sigma(\kappa)$ of the most unstable branch of the dispersion relation for the regimes at $L = 40$ m, $T_0 = 450$ K, $T^0 = 467.46$ K, $P_0 = 10^6$ Pa, $P^0 = 1.19 \cdot 10^6$ Pa. Corresponds to Fig. 6d.

In this case the most unstable perturbations are characterized by small length scale. The formation of the counterflow zone as a result of the annihilation of the stable regime with a sharp interface is one of the possible scenarios of the development of the system after the transition.

5.3. Onset of Instability at $\kappa = 0$

For regimes with $H > 1/2$ ($\min[(1-H)/H, 1] = (1-H)/H$) and $d < (1-H)/H$ the transition to instability takes place at $Z_1 = (1-H)/H$ or, as follows from (5.2), for

$$\sigma_2 = \sigma_1 - \frac{1-2H}{H}\left(\frac{\omega_v H}{1-H}\Gamma_4 R - \omega_w \Gamma_0\right) = 0. \tag{5.4}$$

This equality is attained at the zero value of the wavenumber κ. If instability is developed, the end points of the interval of the wavenumbers, affiliated with growing normal modes, diverge from the origin in the (κ, σ)-plane. The domain of the parameters $\sigma_2 < 0$ corresponds to a stable regime and the domain $\sigma_2 > 0$ corresponds to an unstable one. The stable and unstable dispersion curves $\sigma = \sigma(\kappa)$ of the most unstable branch of the dispersion relation for the regime where this type of transition to instability is realized are pictured in Fig. 4. The onset of instability for regimes with $H < 1/2$ ($\max[(1-H)/H, 1] = (1-H)/H$) and $d > (1-H)/H$ takes place completely analogously but at $Z_2 = (1-H)/H$. The last equality is equivalent to (5.4). The transition to instability for these cases is illustrated in Fig 6b, 6d.

This type of the transition is typical when increase of a vertical permeability of the reservoir occurs (see Fig. 4). In this case the length-scale of the most unstable perturbation is large. This transition is also characterized by annihilation of the stable base solution. It seems that the evolving unstable perturbations lead to the finger formation in this case (at least on the weakly nonlinear stage of their evolution).

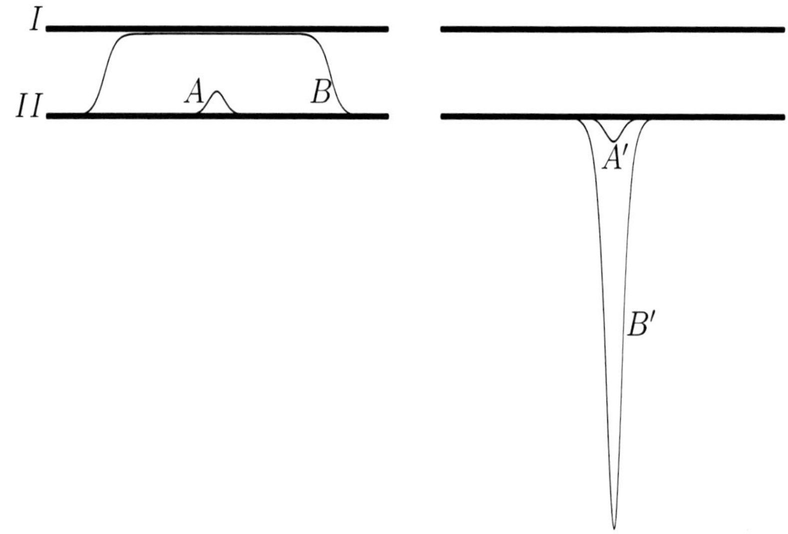

Figure 5. Schematic development of perturbations of the unstable interface II at the threshold of instability. A – the initial perturbation, concentrated between the stable (I) and unstable interface, B – the evolution of A after some time; A' – initial perturbation not entirely concentrated between the stable and unstable interfaces, B' – its evolution after some time.

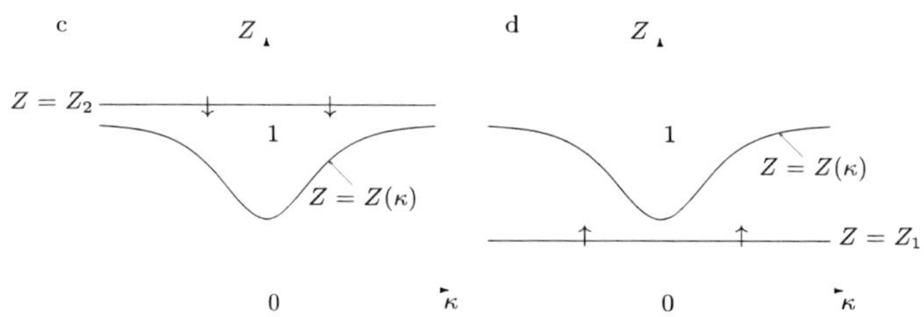

Figure 6. Transition to instability in cases $H < 1/2$ (a,b) and $H > 1/2$ (c,d). The threshold of instability at $\kappa = \pm\infty$ is attained, when either the line $Z = Z_1$ moving upward (a), or the line $Z = Z_2$ moving downward (c), reach the position of the asymptote of the graph $Z = Z(\kappa)$. The threshold of instability at $\kappa = 0$ is attained, when either the line $Z = Z_2$, moving downward (b), or the line $Z = Z_1$ moving upward (d) touch the curve $Z = Z(\kappa)$ at the point $\kappa = 0$.

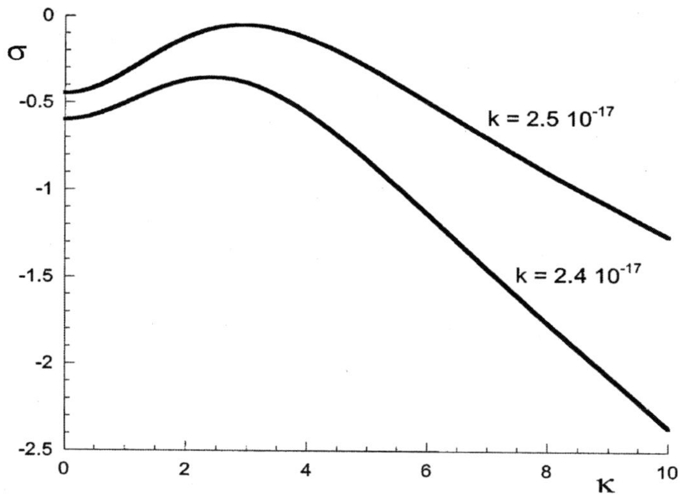

Figure 7. Dispersion curves $\sigma = \sigma(\kappa)$ of the most unstable branch of the dispersion relation for the regimes at $L = 400$ m, $T_0 = 450$ K, $T^0 = 538.96$ K, $P_0 = 10^6$ Pa, $P^0 = 3.62 \cdot 10^6$ Pa ($h = 300$ m). Corresponds to Fig. 8b.

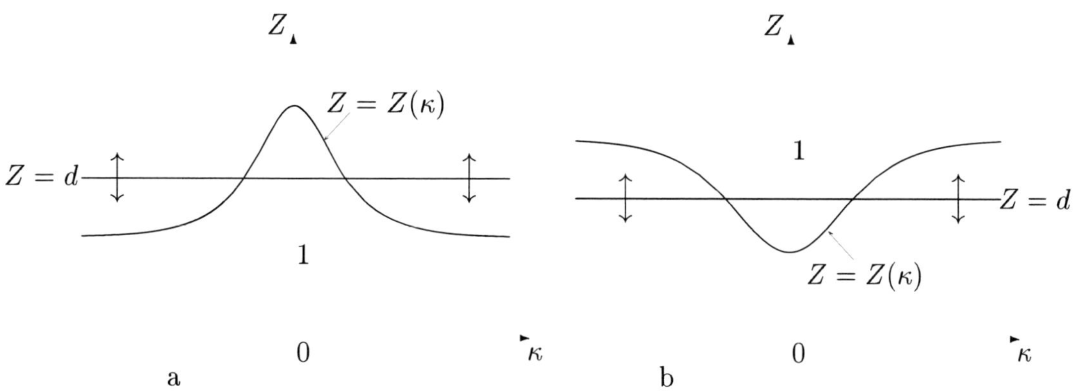

Figure 8. Transition to instability for D in zero neighborhood; $H < 1/2$ (a) and $H > 1/2$ (b). For $D = 0$ the line $Z = d$ arises, which is subjected to disintegration into two lines moving in the opposite directions when the instability is in progress.

As it is underlined in Appendix, the direction of growth of perturbations of the unstable homogeneous base state, corresponding in our physical situation to the unstable vertical flow with plane phase transition interface, depends of the comparative location of the stable and unstable homogeneous states in a neighborhood of the threshold of instability. In our example the unstable interface locates deeper than the stable one, therefore its perturbations grow unboundedly downwards, i. e., there is a tendency for fingers of water to penetrate in the domain occupied by vapor (see Fig 5).

5.4. Onset of Instability at $\kappa \neq 0$

Now, let us consider the case $D = 0$, $\min[(1-H)/H, 1] < d < \max[(1-H)/H, 1]$ of the onset of instability. The equality $D = 0$ is attained at the two non-zero points κ, symmetric about zero. Each of these points gives the double root of the equation (5.2). Crossing the hypersurface in the space of parameters, determined by this equality, the system loses its stability: at $\sigma_1 < 0$ the domain of parameters $D < 0$ corresponds to a stable regime, and the domain $D > 0$ corresponds to a non-stable one. For $D < 0$, the equation (5.2) has no real roots. At $D = 0$ the transition to instability takes place and the roots of (5.2) come to the real axis from the complex plane. In this case the contact of the branch of the dispersion relation with the axis $\sigma = 0$ occurs in a pair of non-zero wave numbers. If variation of parameters leads to the positivity of the value of D, two symmetric about the origin real segments of wave numbers arise, bounded by the roots of (5.2). These wavenumbers are affiliated with unstable normal modes. The stable and unstable dispersion curves $\sigma = \sigma(\kappa)$ of the most unstable branch of the dispersion relation for the regime where this type of transition to instability is realized are pictured in Fig. 7.

This type of instability is typical for reservoirs of the large depth and was considered in the relative problems of the instability in [16]. As it was noted in Introduction the equation governing the evolution of weakly nonlinear perturbation in a narrow band of weakly unstable modes is the Ginzburg-Landau equation. This type of equation describes the regular structures arising as a result of the development of instability.

The transition to instability itself is illustrated in fig. 7a,b.

6. Conclusion and Discussion

In this work we establish the following mechanisms of transition to instability of the interface between the domains occupied by water and vapor in high-temperature geothermal reservoirs.

1. Transition to instability realized in such a way, that all normal modes occur to be growing. In this setting the margin of stability $\sigma = 0$ of the system is attained at all κ and all normal modes become growing simultaneously. This mechanism of transition to instability is realized in the degenerate case $H = 1/2$ of phase transition interface location;

2. $D > 0$; transition to instability takes place through the infinite wave number, i. e.

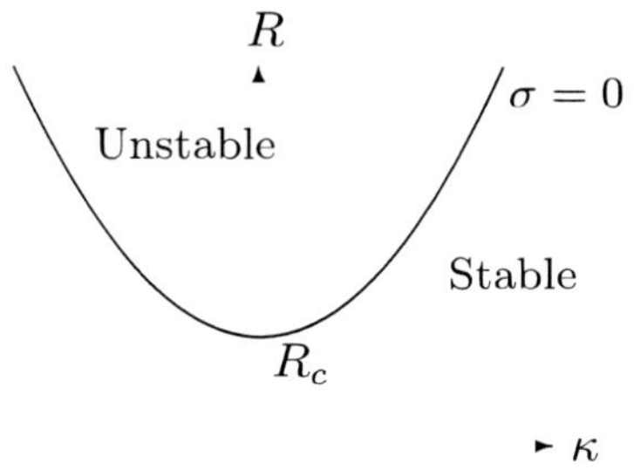

Figure 9. Typical marginal curve $\sigma = 0$ in the (κ, R)-plane.

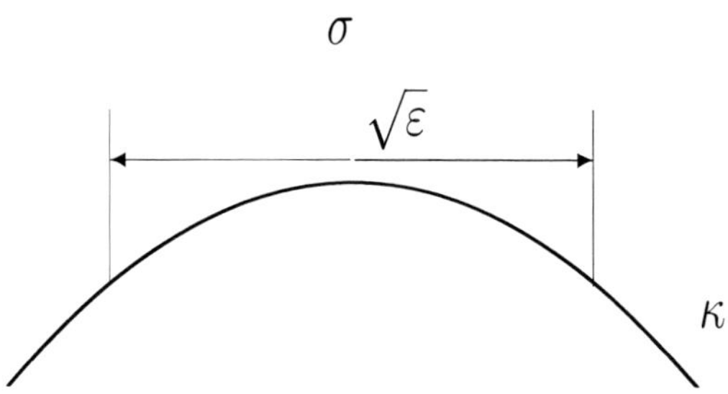

Figure 10. Sketch of the graph $\sigma(\kappa, R)$ for fixed R, $0 < \varepsilon = R - R_c \ll 1$.

the neutrally stable normal modes, corresponding to a threshold of instability appear to be modes, affiliated to the wavenumber $\kappa = \pm\infty$. Such a mechanism of transition to instability is realized at $\sigma_1 = 0$ either if $d < 1$ for $H < 1/2$, or if $d > 1$ for $H > 1/2$ (see (5.3), (5.2)). The system is stable at $\sigma_1 < 0$;

3. $D > 0$; the threshold of instability is attained for $\sigma_2 = 0$ (see (5.4)) at $\kappa = 0$. It takes place either for $H > 1/2$, if $d < (1 - H)/H$, or for $H < 1/2$, if $d > (1 - H)/H$. The annihilated regime evolves into non-stationary nonhomogeneous structure;

4. the threshold of instability is attained for $D = 0$ and $\min[(1 - H)/H, 1] < d < \max[(1 - H)/H, 1]$ at the pair of double non-zero values of the wavenumber κ. At $\sigma_1 < 0$ the domain of parameters $D < 0$ corresponds to a stable regime, and the domain $D > 0$ corresponds to a non-stable one.

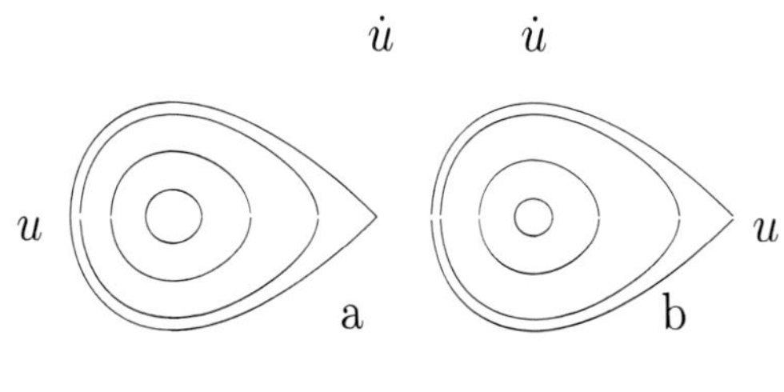

Figure 11. Phase portrait (bounded solutions) of the equation (A4): a –subcritical case; b – supercritical case.

We consider the particular model problem for two-layered geothermal systems, yet we conjecture that the qualitative results of the paper reflect the main features of the dynamics of a general enough two-layered geothermal reservoirs in some neighborhood of the transition to instability.

The case **4** of the transition was considered for the first time in [16]. This type is typical for the reservoirs with a considerably large thickness. The transition is accompanied by the branching of the destabilized vertical regime and caused by the increase of the permeability.

The cases **2** and **3** of the transition were not considered elsewhere to our knowledge before. The typical reason for the transition **2** is the decrease of the reservoir temperature or pressure. The transition in the case **3** is caused typically by the increase of vertical rock permeability in reservoirs of a moderate thickness. Due to the small scales of the most unstable normal modes the plane interface subjected to the transition of the type **2** is a good candidate of the regime giving rise to the counterflow zone after annihilation. On the contrary the scale of the most unstable perturbations for the type **3** of the transition is large. It may mean that the transition leads to the formation of fingers promptly having the character of fingers of water penetrating in vapor domain.

In this paper we considered only non-oscillatory instabilities, when time exponents σ, affiliated with growing normal modes remain real after the exchange of stabilities takes place. We are not still able to prove that there is no exchange of stabilities other than the non-oscillatory one, though all the numerous examples we ever treated give a strong support to this conclusion.

6.1. Appendix. KPP Equation

Let us consider one-dimensional homogeneous motion of a viscous fluid, when the independent spacial and temporal variables are x and t (generalization to the case of higher dimensions is straightforward). Let the corresponding linearized hydrodynamic problem admit normal mode solution $\mathbf{A}e^{(i\kappa x+\sigma t)}$ with real κ and a dispersion relation $\mathcal{D}(\kappa, \sigma, R) = 0$, where R is the parameter of the problem, its variation leads to the loss of stability of the flow in question. We assume that the problem is invariant under the spacial reflection $x \to -x$ and also the validity of the principle of exchange of stabilities, i. e. that $\operatorname{Im}\sigma = 0$ at the origin of instability and $\sigma = 0$ corresponds to the marginal mode.

It follows from the spacial reversibility that the expression $\mathbf{A}e^{(-i\kappa x+\sigma t)}$ is also a solution of the linearized problem, i. e. $\mathcal{D}(\kappa, \sigma, R) = \mathcal{D}(-\kappa, \sigma, R)$ for an arbitrary σ. Assume further that at the margin of instability the most unstable mode is given by the mode with a wavenumber $\kappa(R_c) = 0$, where R_c is a critical value of the parameter. Then, in the generic case algebraic approximation of the dispersion relation for $|R - R_c| \ll 1$, describing the narrow band of weakly unstable modes, has the form

$$\sigma = a - b\kappa^2 + O(\kappa^4), \quad \kappa \to 0, \tag{A1}$$

where $a = r(R - R_c)$ $(r > 0)$, $0 < b \to b_c$ at $R \to R_c$. The marginal curve for the case in question is schematically plotted in Fig. 9. The width of the band of unstable modes has the order $O(\varepsilon^{1/2})$, where $\varepsilon = |R - R_c|$ (Fig. 10).

For small ε the problem admits the approximate solutions having the form $\varepsilon A(X, T)$, where A is a real function, X, T are slow coordinate and time, respectively. In the case of the non-degenerate quadratic nonlinearity one has $X = \sqrt{\varepsilon}x$, $T = \varepsilon t$. According to (A1), the amplitude of the most unstable mode obeys the equation

$$\frac{\partial A}{\partial T} = \frac{a}{\varepsilon}A + b_c\frac{\partial^2 A}{\partial X^2} + \delta\, A^2, \tag{A2}$$

where $\delta \neq 0$ is a coefficient at the nonlinearity. The equation $(A2)$ describes the evolution of the band of weakly unstable long waves in a neighborhood of the threshold of instability. By the scaling transfomation the equation is reduced to the form

$$\frac{\partial u}{\partial \tau} = \gamma u - u^2 + \frac{\partial^2 u}{\partial \xi^2}, \quad \gamma = \mp 1. \tag{A3}$$

In the supercritical case ($\gamma = 1$) the equation (A3) is the well-known Kolmogorov-Petrovsky-Piskunov (KPP) eqnation [11]. In the subcritical case ($\gamma = -1$) the equation (A3) is reduced to the KPP form by the transformation $u \to u - 1$.

Fig. 11a,b illustrates the phase portraits of the equation

$$\ddot{u} = -\gamma u + u^2, \tag{A4}$$

describing the stationary regimes in a neighborhood of threshold of instability for $\gamma = \mp 1$, respectively (dot denotes differentiation with respect to ξ).

The figure shows that in the neighborhood of the threshold of instability there are exactly two points of equilibria of the dynamical system which is equivalent to (A4). These points correspond to homogeneous flows. In subcritical domain ($\gamma = -1$) the solution $u = 0$ is stable, and the solution $u = -1$ is unstable. In supercritical domain the solution $u = 0$ is unstable, and the solution $u = 1$ is stable. Therefore at the crossing of the threshold of instability the general shift of equilibria points describing the homogeneous flows takes place. This shift equals to unity for the dimensionless equation (A3). It can be easily verified that the exponential growth of the positive perturbation of the unstable state at the linear stage of the development of instability is suppressed by the nonlinearity, and as a result the dynamically unstable equilibrium slowly evolves to the stable equilibrium, corresponding to the stable homogeneous flow (see Fig. 5). Conversely, the non-positive perturbations of the unstable uniform state rapidly grow with time to the domain left to the unstable point of equilibrium.

In conclusion we underline the peculiarities of weakly nonlinear development of perturbations of homogeneous flows for the type of transition under consideration. Evolution of perturbations in a neighborhood of the threshold of instability is governed by the amplitude equation (A2), which is reduced by scaling to the KPP equation (A3). There are two states of equilibria in this neighborhood, one is stable and the other is unstable. The positive perturbation of the unstable point of two equilibria of the equation (A3) evolves asymptotically in time to the stable point of equilibrium, corresponding to the respective stable homogeneous regime. The perturbations of the different polarity are subjected to unbounded growth also on the nonlinear stage of their evolution. The comparative location of the stable and unstable states of equilibria depending on the sign of the coefficient δ at the quadratic nonlinearity in (A2) is of particular meaning. This location determines the direction of the growth of perturbations which are not concentrated entirely between the states of equilibria.

This work is supported by Russian Foundation for Basic Research Grant No. 05-01-00554a.

References

[1] Bau, H.H.; Torrance, K.E. *Int J Heat Mass Trans.* 1982, 25, 45–55.

[2] Chandrasekhar, S. *Hydrodynamic and Hydromagnetic Stability*, Oxford University Press: New York, 1967.

[3] DiPrima, R.C.; Eckhaus, W.; Segel, L.A. *J Fluid Mech.* 1971, 4, 705–744.

[4] Drazin, P. *Introduction to Hydrodynamic Stability*. Cambridge University Press: Cambridge, 2002.

[5] Eckhaus, W. In *ICIAM-91, Proc. Second Int. Conf. on Industrial and Applied Mathematics*, 1991.

[6] Grant, M.A. *Geothermics.* 1983, 12, 251–263.

[7] Fitzgerald, S.D.; Woods, A.W. *Nature.* 1994, 367, 450–453.

[8] Hochstein, M.P. In *Symposium on the Development and Use of Geothermal Resources*, **2**, San Francisco, 1975.

[9] Il'ichev, A.T.; Tsypkin, G.G. *Eur J Mech/B Fluids.*, 2005, 24, 491–501.

[10] Il'ichev, A.; Tsypkin, G. To appear in *Proceedings of IUTAM Symposium on Hamiltonian Dynamics, Vortex Structures, Turbulence, August 25-20*, Moscow, 2006.

[11] Kolmogorov, A.; Petrovsky, I.; Piscounov, N. *Moscow Univ Math Bull.*, 1937, 1, 1–25.

[12] McGuiness, M. *Int J Heat Mass Trans.* 1996, 39, 259–274.

[13] Newell, A.C.; Whitehead, J.A. *J Fluid Mech.*, 1969, 38, 279-303.

[14] Pestov, I. (1998) *J. Fluid Mech.* 1998, 364, 273–295.

[15] O'Sullivan, M.J. *Int. J. Energy Res.* 1985, 9, 319–332.

[16] Schubert, G.; Straus, J.M. *J Geoph Res.* 1980 85, 6505–6512.

[17] Truesdell, A.H.; White, D.E. *Geothermics.* 1973, 2, 154–173.

[18] Tsypkin, G.G. In *Proceedings of the 22-nd Workshop on Geothermal Reservoir Eng. Stanford University* January 27-29, Stanford, 1997; pp 483–490.

[19] Tsypkin G.; Il'ichev, A. *Transp Porous Media.*, 2004, 55, 183–199.

[20] Udell, K.S. (1985) *Int J Heat Mass Trans.* 1985, 28, 485–495.

[21] White, D.E.; Muffler, L.J.P.; Truesdell, A.H. *Econ Geol*, 1971, 66, 75–97.

[22] Woods, A.W.; Green, K. In *Proceedings, Twenty-Fifth Workshop on Geothermal Reservoir Eng. Stanford University* January 24-26, Stanford, 2000.

In: Geothermal Energy Research Trends
Editor: Herman I. Ueckermann, pp. 109-119
ISBN: 1-60021-683-8

Chapter 5

SEDIMENTARY CHARACTERISTICS OF COAL BEDS IN INTRAMONTANE BASINS (MASSIF CENTRAL, FRANCE)

Wang Hua [*] ***and Xiao Jun***
Faculty of Earth Resource, China University of Geosciences,
Wuhan, 430074, China

Abstract

Abundant genetic and sedimentary indicators has been found in the thick coal beds from three fault-controlled coal basins on the Central Massif France. A new formation model for thick continental (intra-mountainous) lacustrine peat swap is proposed. In the new coal accumulation mechanism, thick coal beds were associated with various gravity-influenced breccia and sandstone interlayer sediments and the subaquatic gravitary current transported the organic (peat) and inorganic clasts formed in lakeshore swamp were formed in active clastic environment, and were associated with various gravity-influenced mudstone and sandstone interlayers. The presence of a great number of gravity-flow sediments such as detrital flow, diluted slurry flow or turbidity-current sediments in the coal seams, and that of the contemporaneous gravity slump and deformation structures in the coal seams both indicate that the accumulation of the thick coal beds was characterized by the relatively deep water environment and allochthonous sedimentation. This new model interprets reasonably the accumulation mechanism of the thick coal beds developed in the fault basins in the Central Massif (France) and provides a completely new idea with respect to the traditional coal formation models.

Keywords: coal bed, accumulation mechanism, sedimentary characteristic, Central Massif, France.

---

[*] E-mail address: Wanghua1@cug.edu.cn, Fax: (86)-027-87481259; (86)-027-87436746.

Wang Hua, male, born in 1964 in Heilongjiang province (China), received Ph.D from Dijon University (France) in 1991; 1993-1997: taught subjects of geosciences in Mali University (Mali, Africa); professor in China University of Geosciences (Wuhan) from 1998 and active member of AAPG from 1999. His research interests focus on the coal and petroleum geology.

Many coal geologists have studied the sedimentary environments and accumulation of peat (coal) in continental (intra-mountainous) faulted basins□Teichmüller and Teichmüller, 1982; Courel et al., 1986; Vetter, 1986; Bonijoly and Castaing, 1987; Yang, 1987; Li, 1988; McCabe, 1992; Wu et al., 1992; Wang and Courel, 1994; Wu et al., 1997; Wang, 1997; Hu et al., 1998□and they concentrated on the study of accumulation mechanism for thick coal beds (Courel et al., 1986; Wang, 1988; Flores, 1989; Langenberg et al., 1989; Peng et al., 1994; Esterle and Ferm, 1994; Wu et al., 1996; Wang, 1997; Wang et al., 1999; Zhang et al., 1999). The accumulation mechanism for extra-thick coal beds (more than 60m thick) especially from the Fuxin and the Fushun basins (Wu et al., 1997) has not been paid much attention. He suggested a mixed allochthonous - para-allochthonous accumulation mechanism for the extra-thick coal beds influenced by storm and subaquatic gravity flows in continental faulted basins, based on the study of components and textures of coal beds from these basins in Eastern China.

In this study, the strata outcropping in three small continental intramontane faulted basins in France are studied in detail and new knowledge about the origin of coal beds formed in faulted basins is obtained (Wang and Courel, 1993; Wu et al., 1996; Wang, 1997).

1. Characteristics of Coal Beds in Faulted Basins, France

There are about 20 small faulted basins developed on the Central Massif, France. These basins were formed in Late Carboniferous-Permian time. The three faulted basins (Fig. 1) studied are the La Machine basin (less than $50km^2$), and the Cevennes (less than $140km^2$) and Montceau-Les-Mines basins (less than $150km^2$ (Bordonne et al., 1986).

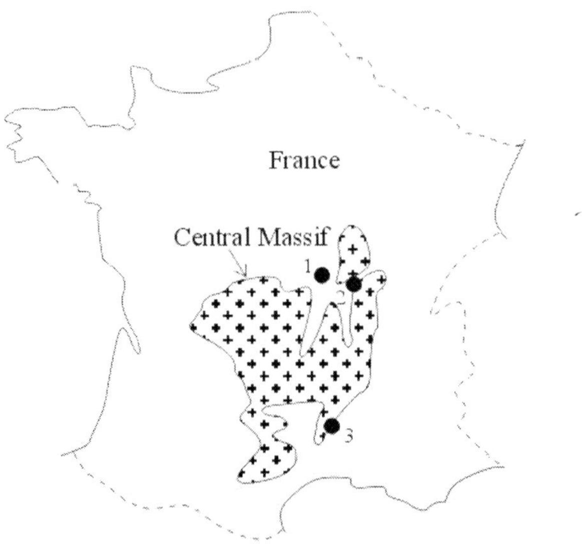

Figure 1. Location of the three coal basins in the Central Massif, France (after Courel et al., 1986). 1. La Machine basin; 2. Montceau-Les-Mines basin; 3. Cevennes basin..

1.1. Characteristics of Coal Beds in Cevennes Basin

The Late Carboniferous Cevennes basin is located in the southeastern margin of the Central Massif, France (Fig. 1 and Fig. 2d). The basement rocks are composed mainly of metamorphosed mica-gneiss, felsic granulite and minor granite (Gras, 1970; Frere, 1984; Faure and Becq-Giraudon, 1993; Disnar et al., 1997). The overlaying coal-bearing series is nearly 2500m thick, and consists mainly of conglomerate, sandstone, siltstone, mudstone and coal beds.

The thickness of the major coal bed, for example, the Sans-Nom coal bed (with low ash matter, < 10%), is up to 40m (Fig. 2a). The thickness of the seam is strictly controlled by the synsedimentary faults in the basin (Wang and Courel, 1993; Wang, 1997; Djarar et al., 1997). Abundant wrinkled, fractured and mixed sedimentary structures related to gravity sliding can be observed in the coal bed (Fig. 2b , 2c and Fig. 3). coal basin (after Djarar et al., 1997). a. Relationship of coal bed to sandstone body; b. Gravity- influenced slump structure in coal bed; c. Distribution of strata and coal bed at two sides of the Malperthus gravity decollement fault; d. Location of Cevennes basin.

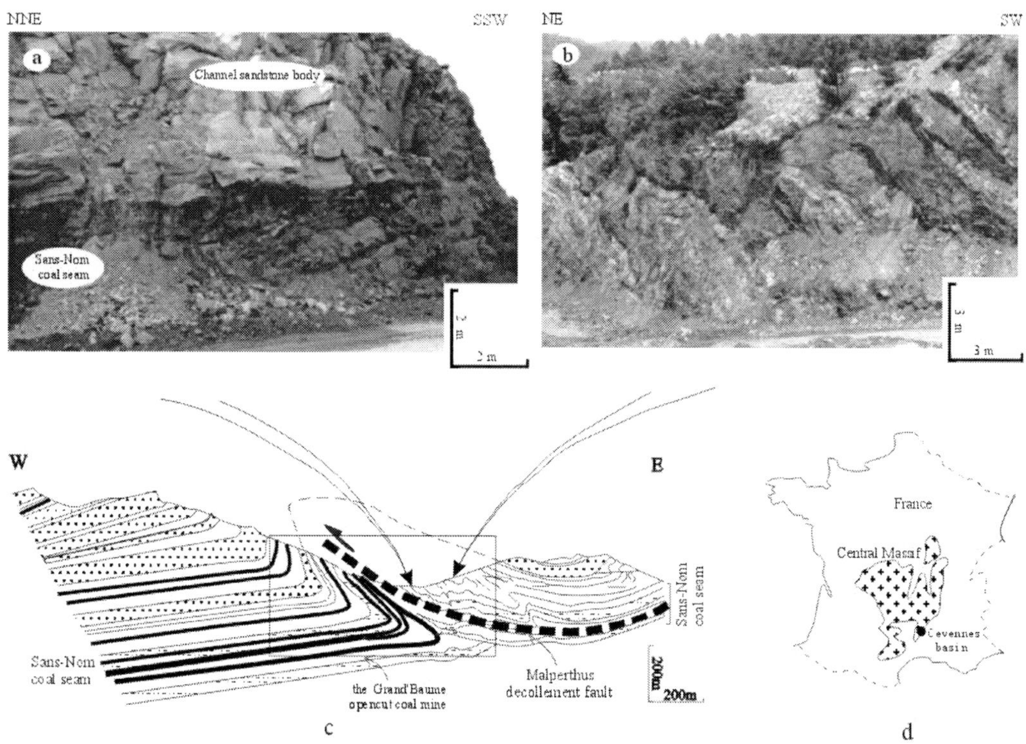

Figure 2. Characteristics and sedimentary setting of the Sans-Nom thick coal seam in the Cevennes.

Figure 2. Caractéristiques du dépôt de la couche de charbon épaisse «Sans nom» dans le bassin des Cévennes (d'après Djarar et al. 1997). a- relations entre charbon et grès; b- structure gravitaire: slump dans le charbon; c- distribution des couches et stériles de part et d'autre de la faille de Malperthus (faille de décollement); d- localisation du bassin des Cévennes.of the Malperthus gravity decollement fault; d. Location of Cevennes basin.

Figure 3a shows two layers of breccia composed of sandstone, conglomerate, siltstone and coal. The fragments have sharp edges, and are different in size, with a maximum diameter up to 40cm. The coal bed is in sharp contact with the underlying feldspar-quartz sandstone, and no rooted clay is found. The coal bed is also in sharp contact with the overlying fractured sandstone with sharply-defined erosion surfaces (Fig. 3a and 3b). In some places the coal bed penetrates into the overlying sandstone, resulting in a “flame structure” (Fig. 3b). Gravity gliding structures can be seen on the Grand’Baume opencut coal mine (Djarar et al., 1997; Fig. 2c).

The coal measures were formed in active stretch tectonic setting (Delenin et al., 1988), and gravity slump structures, which are extremely abundant in the sedimentary-filling sequences (Wang, 1997; Djarar et al., 1997).

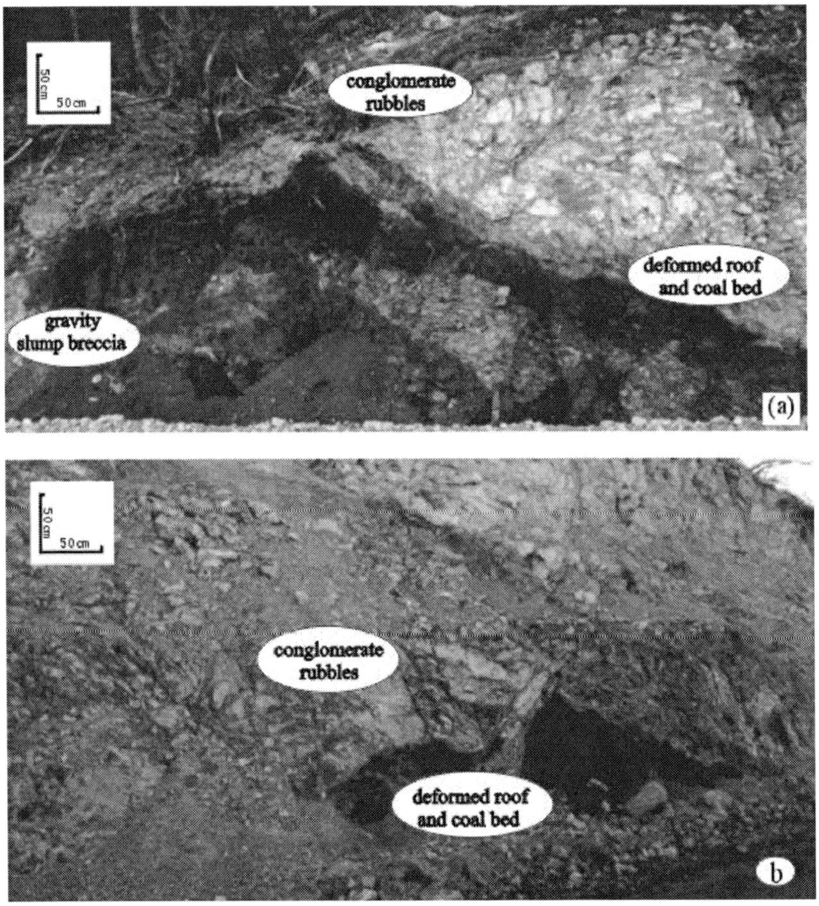

Figure 3. Sedimentary characteristics and architectures of the Sans-Nom coal seam in the Late Carboniferous Cevennes basin, France. a. Gravity slump breccia in thick coal seam; b. Intensely deformed roof and coal bed with numerous irregularly lenticular sandstone and conglomerate rubbles.

1.2. Characteristics of Coal Beds in Montceau-Les-Mines Basin

The Montceau-Les-Mines coal basin is located in the northeastern margin of the

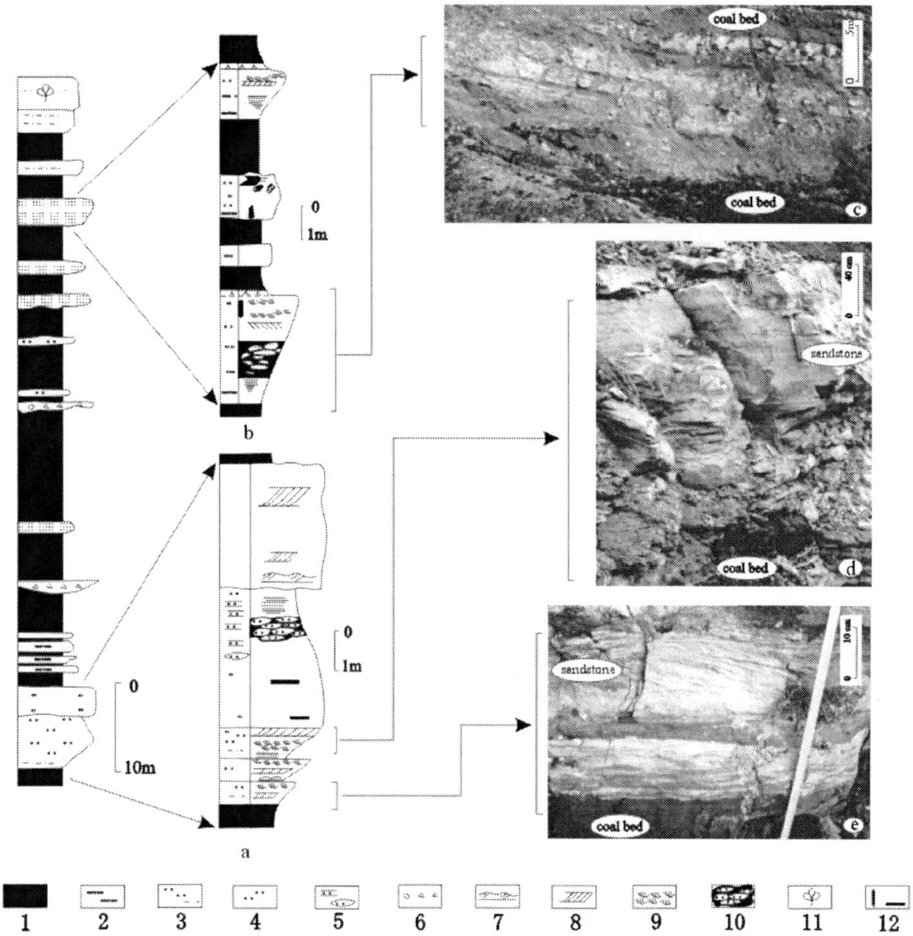

Figure 4. Sedimentary sequences and architecture characteristics in the middle and upper part of the No. 1 thick coal seam from the Montceau-Les-Mines basin, France. a and b: coal bed columnar sections and sedimentary structures; c, d and e: sandstone bodies with reverse-graded bedding in the No.1 thick coal seam. 1- coal seam; 2-mudstone; 3-siltite; 4-sandstone; 5-sandstone bed and lenticular body; 6-conglomerate; 7-ripple mark; 8-tabular cross-bedding; 9-trough cross-bedding; 10-lenticular bedding; 11-plant leaf prints; 12-plant column fossil.

Central Massif (2 in Fig. 1), and is also a small continental faulted basin (Wang, 1988; Valle et al., 1988; Golitsyn et al., 1997), which strikes along a NE-SW direction. The basement is composed of granitoid rocks. The overlaying coal measures is nearly 2000m thick, and is composed of granitic conglomerate, lithic sandstone, arkose, black gray mudstone and coal beds (Fig. 4a and 4b). There are 20 minable seams in the coal-bearing series, among which the No. 1 seam is the thickest, up to 100m (ash matter < 9%).

The small lenticular sandstones developed in the middle and upper part of the coal series, are mainly composed of moderate- to fine-grained sandstone and siltstone, and show reverse-graded bedding (Fig. 4c, 4d and 4e).

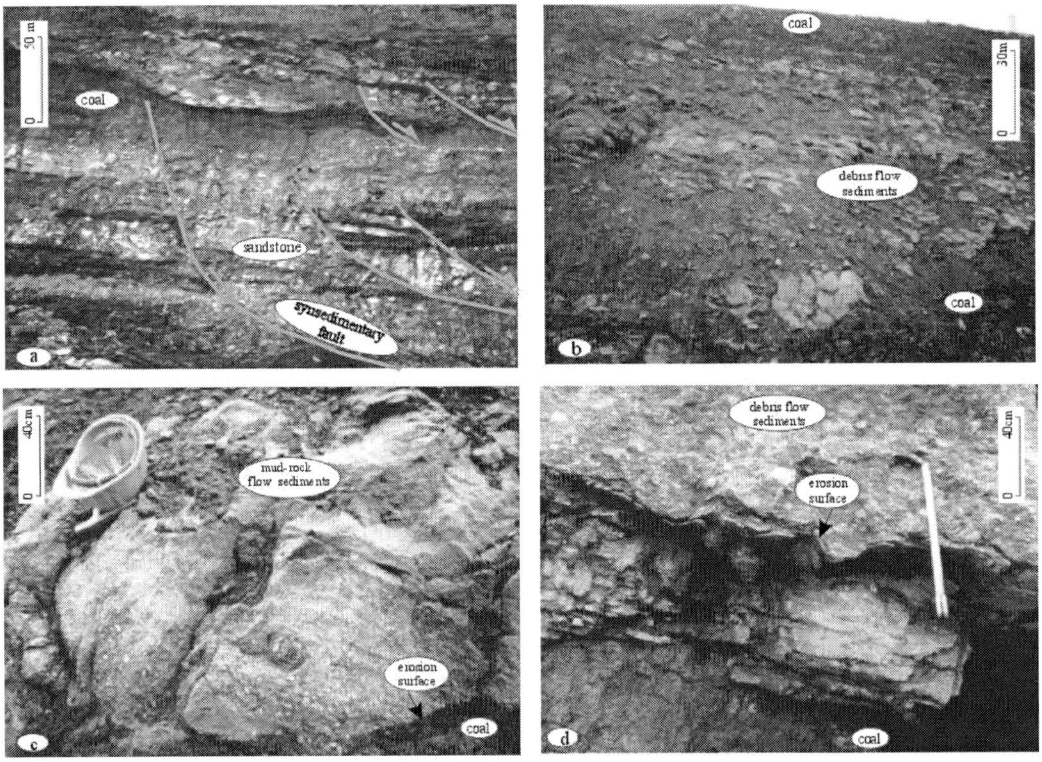

Figure 5. Sedimentary architectures of the No.1 thick coal seam in the Late Carbonifcrous Montceau-Les-Mines basin, France. a. Types of syndepositional structure and the features of sedimentary bodies; b. Abundant gravity-influenced clastic current sediments in extra-thick coal bed; c. Lenticular rare mud-rock flow sediments in extra-thick coal bed; d. Mud-rock flow sediments with scouring surface in extra-thick coal bed.

In the No. 1 seam are found abundant lenticular breccia partings, thin-bedded and lenticular sandstones (Fig.5a, 5b and 5c) and one layer of volcanic tuff. The thin-bedded sandstones with a poorly-defined graded bedding. The bottom of the seam contains a great amount of crystalline pyrite, and show ripple, small-scale horizontal bedding, convolute bedding and vein bedding.

The lenticular breccias are developed in the middle-lower part of the coal series, all with a convex base and flat top (Fig. 5c and 5d). Single layer of lenticular breccia is 1-2m thick. The breccia shows matrix support texture and poor sorting. The rubbles are dominated by granite, gneiss and quartz and feldspar fragments, with a maximum diameter up to 50cm, while the matrix is composed of sandstone, mudstone and coal debris (with a maximum diameter up to 2cm). In the middle and upper parts of the breccias are para-bedded or flattened lenticular coal seams or coal streaks (Wang, 1988). The breccias show a poorly-defined fining-upward sequences that contains siltstone and mudstone. The overlying coarse-grained lithic sandstones have a moderate sorting and grain- and matrix-supported texture, but with a massive appearance. The breccias cause intense scouring on the underlying seams or mudstone, and resulted in the formation of diapiric structure and load structures (Fig. 5c and 5d), producing obviously sharp contact between the breccias and underlying seam. Also intense deformation and fold structures can be seen in seams.

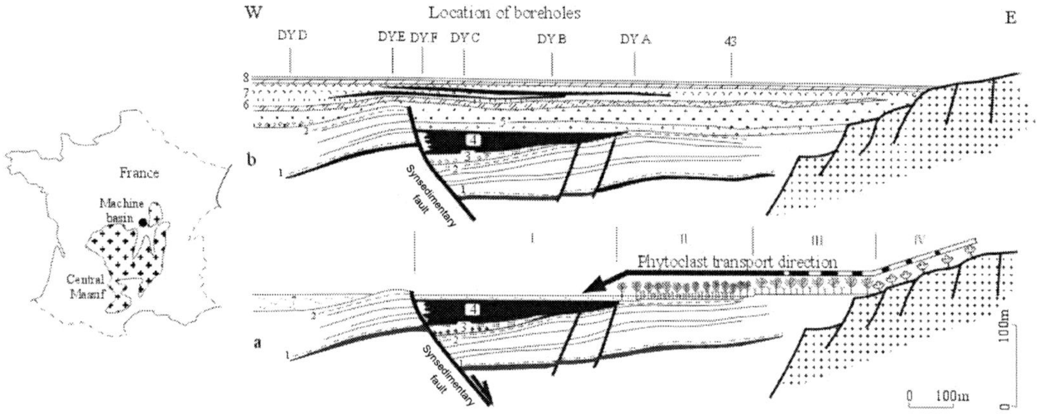

Figure 6. Accumulation model and sedimentary environments of the thick Les Varioux coal bed from the Machine coal basin, France (adapted from Donsimony,1983; see Fig.1 for location). 1 and 2. volcanic tuff; 3. sandy conglomerate in the lower part of the thick Les Varioux coal bed; 4.thick Les Varioux coal bed; 5. polymictic breccia and conglomerate; 6. clay rock; 7. clastic sediments and coal streak and seam; 8. stromatolite; I. water body; II. growth area of hydrophilous plant; III. intermittent swamp IV. growth area of xerophyte.

The sandstone body has a maximum thickness up to 90m and width of 120-150m (Fig. 5b). The sandstone body, from bottom to top consists of coarse pebble-bearing sandstone showing matrix-supported texture and poor sorting, moderate- to fine-grained sandstone and siltstone with coal debris. The sandstone body contains abundant organic materials, is dark in color, and shows poorly-developed sedimentary structure (Wang, 1988) . It can be deduced that the sandstone body may be formed by mud-rock flow.

The sedimentary sequence at the bottom of the coal series is mainly composed of sandstone lenses, and shows either normal graded bedding or inverse graded bedding, or no graded bedding (Fig. 5b and 5d). Laterally, the sandstone bodies are very discontinuous (Fig. 5c). The middle part of the coal series is characterized by the well-developed breccia bodies resulted from gravity debris flow (Fig. 5). The thick immature sandstone lenses was formed by mud-rock flow (Fig. 5d). The upper and top parts are characterized by well-bedded sandstone and mudstone with typical reserse graded bedding (Fig. 4). In some places, the sandstone body, however, is displaced by a syndepositional fault. Close to the fault the sandstone has a considerable variation in thickness and constitutes several small foreset sandstone bodies (Fig. 5a). Vertically, the thick No. 1 seam shows a grading variation of coarse-fine-coarse from bottom to upper, with a normal graded bedding at the bottom and a reverse graded bedding at the top. The seam is characterized by gravity flow sediments, and slump and wrinkled structures.

1.3. Sedimentary Characteristics of Thick Various Coal Beds in La Machine Basin

The La Machine basin is located in the northern margin of the Central Massif (Fig.1 and 6). The main Varioux coal bed is up to 39.5m thick and only 450m long in east-west direction. This seam is controlled by a N-S trending syndepositional normal fault, and is in angular discordance contact with the underlying strata (Fig.6a and 6b). The floor, in which no rooted clay are found, is made of breccia (3 in Fig.6). Generally, the roof is made of polymictic breccia, and is in sharp, erosion or scouring contact with the seams. Donsimony (1983) and Courel. et al. (1986) suggested that the seams were formed in deep-water lakes where no well-developed plant community grew. Plants growing in shallow areas were transported into lakes to form peat. The transportation and accumulation way - typical allochthonous accumulation model is shown in figure 6.

2. Discussion and Conclusions

The studies on the abundant origin traces found in the thick coal beds from the three small faulted basins indicate the following: (1) The above-mentioned features suggest that the coal beds was accumulated in a high-energy, deep-water environment, in which were formed also subaqueous sheet-like sandstone bodies, mud-rock sediments, debris flow sediments, distributary channel sandstone bodies (Wang, 1988; Wang et al., 1994; Courel et al., 1994) . The lithofacies were obviously controlled by margin faults and syndepositional faults and which were considerably variable in relative elevation and were close to provenance (Courel, 1989; Genna and Debriette, 1996). (2) Extremely abundant gravity-flow sediments and rare mud-rock-flow sediments in coal beds show that the coal beds were transported into deep-water environments (>10m) but broad shallow-water warms (<3m). (3) The abundant normal and deformed structures, as convolute bedding, slump, fractured structures found in the coal beds imply that peat was transported by the gravity flows and deposited in unstable environments.

Although the three basins are separated by a very long distance and they were formed in different geologic ages. Therefore, all these basins contain major (thick) coal beds and the same sedimentary bodies. These basins were formed especially in active tectonic setting, with intensely active marginal and basin faults, which controlled the time and space distribution of sedimentary facies zones and coal-rich units. Numerous gravity current sediments, as well as abundant syn-depositional gravity slump structure, deformed structure and indicators for re-transportation and re-deposition are found in the major coal beds. All these features suggest that the thick coal beds mentioned above were transported in the unstable deep-water environments, that the peat clasts are of allochthonous origin (re-deposited). In fact, thick coal beds in many coal basins often include abundant gravity current sediment interlayers. Traditionally these interlayers and their important sedimentary significance are neglected. The presence of gravity current sediments in coal beds suggests that some peat clasts were accumulated allochthonously in deep-water environments with high energy (with low ash matter, <10%), i.e. active clastic environments, but not low-energy swamps as we known.

Acknowledgments

The authors would like to thank Professor Louis COUREL from Dijon University and geological engineer Mr. Jacques Laversannes from "Charbonnage de France", geological engineer Mr. Michel Dumain from "Houilleres des Cevennes" for their help in field survey and information collection.

References

Bonijoly, D., Castaing, C., 1987, Ouverture et évolution structurale de quelques bassins houillers de directions orthogonales dans le Massif Central français. *Ann. Soc. Geol. Nord,* **6**: 189-200

Bordonne, G., Douay, F., Paquette, Y., 1986, Les conditions de gisements particulières des bassins houillers limniques intra-hercyniens; importance dans le choix des methodes d'exploitation et l'évaluation des ressources. *Mém. Soc. géol. France, N.S.,* **149**: 73-86

Courel, L., Donsomoni, M., Mercier, D., 1986, La place du charbon dans la dynamique des systèmes sédimentaires des bassins houillers intramontagneux. *Mem. Soc. géol. France, N.S.,***149**:37-50

Courel L., 1989, Organics versus clastic: conditions necessary for peat (coal) development. *International Journal of Coal Geology*, **12**: 193-207

Courel, L., Valle, B., Branchet, M., l994. Infilling dynanmics of the intramontane basin of Blanzy-Montceau (Massif Central, France). Int C. Poplin and D. Hey1er (Editors). Quand le Massif Central etait sous l'Equateur. Mém. Section Sci., *Comite des Travaux Historiques et Scientifiques*, **l2**: 33-45

Delenin, P., Clermonte, J., Courel, L. et al., 1988, Remise en cause des charriages dans le bassin houiller Stéphanien des Cévennes (Gard, France). *C. R. Acad. Sci. Paris,* **307**(11):1237-1243

Disnar, J.R., Marquis, F., Espitalie, F. et al.,1997, Organic geochemistry and reconstitution of thermal and tectono-sedimentary history of the Ardeche Margin (GPF program; France). *Bull. Soc. Geol. France,* **168**(1):73-81

Djarar, L., Wang, H., Guiraud, M. et al., 1997, The Cevennes Stephanian basin (Massif Central): an example of relationships between sedimentation and late-orogenic extensive tectonics of the Variscan belt. *Geodynamica Acta* (*Paris*), **9**,5,193-222

Donsimony M., 1983, Le gisement houiller de Decize-Devay (Nièvre, France). Influence d'une tectonique synsédimentaire sur la genèse d'une couche épaisse. *X° congr. Strat.Géol. Carb, Madrid, C.R.***3**, p 333-341.

Esterle, J.3., Ferm, J. S., 1994, Spatial variability modern tropical peat deposits from Sarawak, Malaysia and Sumatra, Indonesia: analogues for coal, *International Journal of Coal Geology.* **26**(1~2):1~41

Faure, M., Becq-Giraudon, J.P., 1993, Sur la succession des episodes extensifs au cours du desepaississement carbonifère du Massif Central francais. *C. R. Acad. Sci. Paris,* **316**(II): 967-973

Flores, R.M., 1989, Rocky mountain tertiary coal-basin model and their applicability to some world basins, *International Journal of Coal Geology,***12**: 767-798

Frere, I., 1984, Le bassin Stéphanien des Cévennes (Gard). Dynamique du remplissage – place du charbon. *Cinérites*. These de 3ème cycle, Univ.Dijon (France); 172pp

Genna, A., Debriette, P.J., 1996, Structural evolution of the Ales coal basin (new data and interpretation). *Bull. Soc. Geol. France.* **167**(1):83-91

Golitsyn, A., Courel, L., Debriette, P., 1997, A fault-relatad coalification anomaly in the Blanzy-Montceau Coal Basin (Massif Central, France). *Int. J. Coal Geol.*,**33**:209-228

Gras, H., 1970, Etude géologique détaillee du bassin houiller des Cévennes (Massif-Central francais). *Mém. Renéot. Houillères des Cévennes.* 305pp

Hu, Y.C., Liao, Y.Z., Li, Z.M., 1998, Late Carbonif alluchthonous coal of Yiluo coalfield in Henan Province (in Chinese), 23(6): 589-594

Langenberg, W., Macdonald, D., Kalkreuth, W. et al., 1989, Coal quality variation in the Cadomin-Luscar coalfield, Alberta: Alberta Research Council, *Earth Sciences Report* **89-**1, 65pp

Li, S.T., 1988, Fault basins analysis and coal accumulation. –an approach to sedimentation, tectonic evolution and energy resource prediction in the late Mesozoic fault basins of Northeastern China. Geology Press of China, Beijing.326pp (in Chinese)

McCabe, P.J., 1992, Tectonic and climatic controls on the distribution and quality of Cretaceous coals. *Geological Society of America, Special Paper* **267**:1-15

Peng, G.L., Zhang, Z.Y., Wu, D.M., 1994, *The formation of thick-bedded peat and modern coal-accumulating processes.* Central South University of Technology Press, Changsha. 112pp (in Chinese)

Teichmüller, M. Teichmüller, R., 1982, The geological basic of coal formation. In: *Stach's Textbook of coal petrography*, ed. by E. Stach, M. –Th. Mackowsky, M. Teichmüller, G.H. Tayer, D. Chandra and R. Teichmüller, Gebruder Borntraeger, Berlin, pp.5-86

Valle, B., Courel, L. Gelard, J.P., 1988, Les marqueurs de la tectonique synsédimentaire et syndiagénétique dans le bassin stéphanien à regime cisaillant de Blanzy-Montceau (Massif Central, France). *Bull. Soc. Géol. France.* **IV**(4):529-540

Vetter, P., 1986 Les formations limniques du Carbonifère supérieur et de l'Autunien en France. *Soc. géol. France, N.S.,***149**: 7-14

Wang, H., 1988, Sédimentation et tectonique précoce dans la 1ère couche de l'Assise de Montceau (Découverte Barrat; Bassin houiller Stéphanien de Blanzy-Montceau). D.E.A. Univ. Dijon (France), p.11-17, 34-43

Wang, H., Courel, L., 1993, Evolution of synsedimentary strctures and analysis of basin tectonic setting. *Earth Sciences*-Journal *of China University of Geosciences.* **18**(2):129-138(in Chinese)

Wang, H., Courel, L., 1994, Block model and their relationship to peat (coal) accumulation in faulted basin of Montceau-Les-Mines, France. *Coal Geology and Exploration* , **22**(5):1-6(in Chinese)

Wang, H., 1997, Analyses géologique du bassin houiller stéphanien des Cévennes (France). *Edition Jamana,* 254pp

Wang, H., Wu, C.L., Courel, L. et al. 1999, Analysis on accumulation mechanism and sedimentary condition of thick coal beds in Sino-French faulted coal basins. *Earth Science Frontiers,* **6** (Suppl.):157-166 (in Chinese)

Wu, C.L., Li S.T., Cheng, S.T., 1992, Humid-type alluvial-fan deposits and associated coal seams in the Lower Cretaceous Haizhou Formation, Fuxin Basin of northeastern China. *Geological Society of America Special Paper* **267**, 269-286

Wu, C.L., Wang, G.F., Li, S.H. et al., 1996, Study on allochthonous genesis of ultra thick coal beds in continental faulted basin, *Geological Science and Technology Information*, **15**(2): 63-67(in Chinese)

Wu, C.L., Li, S.H., Huang, F.M., 1997, Analysis on the sedimentary conditions of ultra-thick coal beds in Fushun basin, *Coal Geology and Exploration*, **25**(2): 1~6 (in Chinese)

Yang, Q., 1987, Advances in coal geology, Science Press, Beijing, China), 316p (in Chinese)

Zhang, R.S., Wu, C.L., Li, S.H., Xu, B.Q., 1999, Accumulation dynamic mechanism of the main coal seam of the eocene Guchengzi Formation in Fushun Coalfield. *Geological Review*, **45**(6): 654-660 (in Chinese)

In: Geothermal Energy Research Trends
Editor: Herman I. Ueckermann, pp. 121-161
ISBN: 1-60021-683-8

Chapter 6

COUPLING OF THERMAL AND CHEMICAL SIMULATIONS IN A 3-D INTEGRATED MAGMA CHAMBER-RESERVOIR MODEL: A NEW GEOTHERMAL ENERGY RESEARCH FRONTIER

Surendra P. Verma* and Jorge Andaverde
Geoenergía, Centro de Investigación en Energía, Universidad Nacional Autónoma de México, Privada Xochicalco s/no., col. Centro, Temixco, Mor. 62580, MEXICO

Abstract

As an innovation, we propose that a new frontier in geothermal research should be explored that involves the coupling of thermal and chemical simulations in an integrated "magma chamber-reservoir" model. To achieve this innovation in geothermal research, we have written a new computer program (in Fortran 90) in modular structure that runs on a PC under the dynamic memory concept and simulates heat transfer conductive and convective processes both in a magma chamber and the overlying geothermal reservoir as well as computes *in-situ* major element chemistry of magmas that evolved in the magma chamber as a result of processes of assimilation, crystallization (liquid line of descent), magma mixing, recharge, and eruption. This combined task is accomplished in three dimensions (3-D) – a substantial improvement as compared to the current practice of obtaining thermal solutions in 1-D or 2-D and of modeling chemical data obtained from the analysis of surface outcrops *without* reference to the actual location within the Earth where the magmas were stored prior to eruption. In fact, if temperature estimates in drill wells and chemical data for surface rocks were available, this information can be used to constraint the model we are proposing as a new research frontier. The practice of "direct" modeling can be replaced in the future by inverse modeling when greater computing and storage capacities of personal computers will be available. This chapter briefly reviews the current state of thermal modeling of geothermal areas and presents the salient features of our new research approach, including a brief description of our computer program. An application example of a Mexican geothermal field (Los Humeros, Puebla), currently under exploitation for electricity production, will highlight the use of our software. This particular geothermal field was chosen for illustration purposes because of the availability of required thermal and chemical data to test the 3-D simulation

* E-mail address: spv@cie.unam.mx (Corresponding author)

model. We have successfully reproduced some of the major element chemical characteristics for the most voluminous caldera-forming eruption at about 0.46 Ma and the present-day thermal regime inferred from static formation temperatures using a quadratic regression of the actually measured bottom hole temperature data.

Introduction

Geothermal energy, being a viable energy resource [Barbier 1985; Guglielminetti 1985; Verma 2002a; DiPippo 2005], has been investigated using conventional geological, geochemical, and geophysical methods. Investigations of geothermal systems are generally split into the initial exploration stage followed by feasibility period and eventual exploitation for electricity production. A conceptual model of the field thus evolves through these investigations.

As an illustrative example of a well studied area, the Long Valley caldera, California, USA, can be mentioned, where a large number of studies from different geoscientific fields have given way to a three-dimensional conceptual model for this field [DiPippo 2005]. These studies include (only to mention a few of them in chronological order): simple heat conduction models [Lachenbruch et al. 1976]; use of chemical and isotopic geothermometers [Fournier et al. 1979]; surface deformation studies [Denlinger and Riley 1984]; magnetotelluric studies [Hermance et al. 1984]; unrest monitoring [Hill 1984]; use of anomalous earthquake signals for location and configuration of magma bodies [Sanders 1984]; earthquake swarms as an evidence for dyke inflation [Savage and Cockerham 1984] and magma chamber studies [Luetgert and Mooney 1985]; geochemistry of volcanic rocks to infer multiple magmatic systems [Sampson 1987; Sampson and Cameron 1987; Vogel et al. 1994]; barometric data from mineral chemistry of volcanic rocks to infer the depth of last magma equilibration [Johnson and Rutherford 1989]; geochemistry, stable isotope geology, and geochronology of rocks [McConnell et al. 1995, 1997; van der Bogaard and Schirnick 1995]; mobility of chemical elements in the hydrothermal system [Wollenberg et al. 1995]; and inversion techniques for inferring the nature and shape of magmatic sources [Tiampo et al. 2000]. Besides these papers, a special issue of the Journal of Volcanology and Geothermal Research was dedicated to "Crustal unrest in the Long Valley caldera, California: New interpretations from geophysical and hydrologic monitoring and deep drilling" [Sorey et al. 2003], which included contributions on: relationships between seismicity and deformation [Hill et al. 2003]; mechanisms of unrest [Battaglia et al. 2003a]; geodetic and microgravity studies [Battaglia et al. 2003b]; deformation studies [Langbein 2003; Howle at al. 2003]; relationships of water-level changes with local and distant earthquakes [Roeloffs et al. 2003]; pumping tests and geochemical modeling [Farrar et al. 2003]; thermal data and deep electrical sounding [Pribnow et al. 2003]; and mineralization studies to infer history of the hydrothermal system [Fischer et al. 2003]. However, in spite of such a large number of studies no combined thermal and chemical modeling in 3-D has yet been reported even from this well-studied caldera area.

In a given geothermal area, once exploitation has begun as a result of fluid extraction from drill wells, the practice has been to intensify research in the field of reservoir engineering, but *without* actually taking into account the primary heat source responsible for the emplacement of such geothermal or hydrothermal systems [e.g., Ferriz, 1982; Economides, 1985; Tan, 1985; Sánchez Upton 1994; Bertani and Cappetti 1995; Meidav

1995; Moya et al., 1995, 1998; Torres R., 1995; DiPippo 2005]. On the other hand, most such systems are emplaced in young volcanic terranes, for example, Los Humeros geothermal field is situated in a caldera in which the volcanic activity took place during the last ~0.47 Ma, with the most voluminous caldera-forming event at ~0.46 Ma and the youngest basaltic magma eruption at <0.02 Ma [Ferriz and Mahood, 1984]. Further, it is important to note that many geothermal fields around the world are located within certain geological structures known as calderas. Such geological structures, therefore, should be the focus for geothermal exploration programs [Anguita et al. 2001].

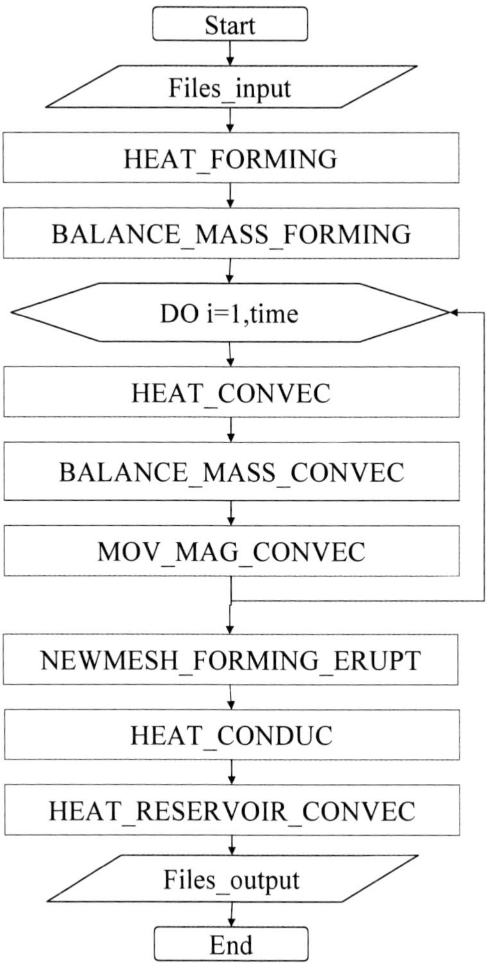

Figure 1. Basic modular structure of TCHEMSYS computer software, written in Fortran 90. More details on different modules can be seen in Figs. 2 and 3.

Geochemical studies involving rock as well as fluid compositions are also carried out independently *without* a proper reference to the thermal constraints of the primary heat source commonly thought to be a "magma chamber", for example, for the Los Humeros geothermal field [Verma and Lopez M., 1982; Verma, 1983, 1984, 2000; Ferriz and Mahood, 1984, 1987; Ferriz 1985], Los Azufres geothermal field [Dobson and Mahood 1985; Torres-Alvarado and Satir 1998; González-Partida et al. 2005; Verma et al. 2005; Pandarinath et al. 2006], La

Primavera geothermal field [Mahood 1981a, b; Mahood et al. 1983; Mahood and Halliday 1988], and Las Tres Vírgenes geothermal field [Verma et al. 2006a].

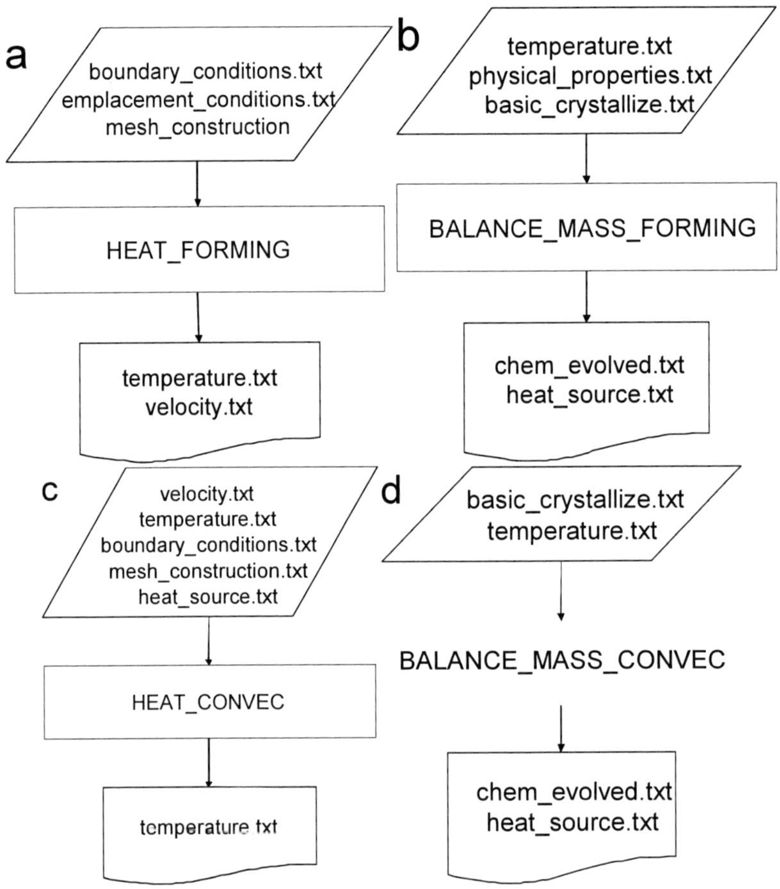

Figure 2. Explanation of the first four programs of TCHEMSYS computer software, with more explicit information on input and output files. (a) HEAT_FORMING; (b) BALANCE_MASS FORMING; (c) HEAT_CONVEC; and (d) BALANCE_MASS_CONVEC.

It is clear that a combined thermal and chemical modeling of magma chambers could be an important step forward to better understand the complexity of geothermal systems. Therefore, as an innovation we propose that a new frontier in geothermal research should be explored that involves the coupling of thermal and chemical simulations in an integrated "magma chamber-reservoir" model. This is precisely what we present in this chapter, with an emphasis on the description of a new computer software, written by us in Fortran 90, to simulate thermal and chemical predictions in 3-D and its application to the Los Humeros geothermal field of Mexico.

Previous Studies

Magmatic intrusions in the form of a magma chamber are probably the primary heat source in most geothermal areas around the world [e.g., Verma 1985; Luhr and Williams 1991;

DePippo 2005]. The overlying reservoir rocks heated from this source serve the purpose of exploitation of geothermal energy to produce electricity when a hydrothermal system is well established, i.e., when, as a result of natural processes, superheated steam or water is available for extraction and exploitation.

The difficult subject of rather complex magma chamber processes has been reviewed by [Cashman and Bergantz 1991; Bergantz 1995], among others. Thermal modeling of a magma chamber and study of geologic processes taking place within the chamber have constituted an advancement in understanding the heat budget of a given area [Spera 1980; Spera et al. 1982; Giberti et al. 1984a, b; Tait 1988; Giberti and Sartoris, 1989; Marsh 1989, 1996; Verma et al. 1990; Valentine 1992; Andaverde et al. 1993; Jaupart and Tait 1995; Verma and Andaverde 1996; Verma and Rodríguez-González 1997; Stimak et al. 2001; Valentine et al. 2002].

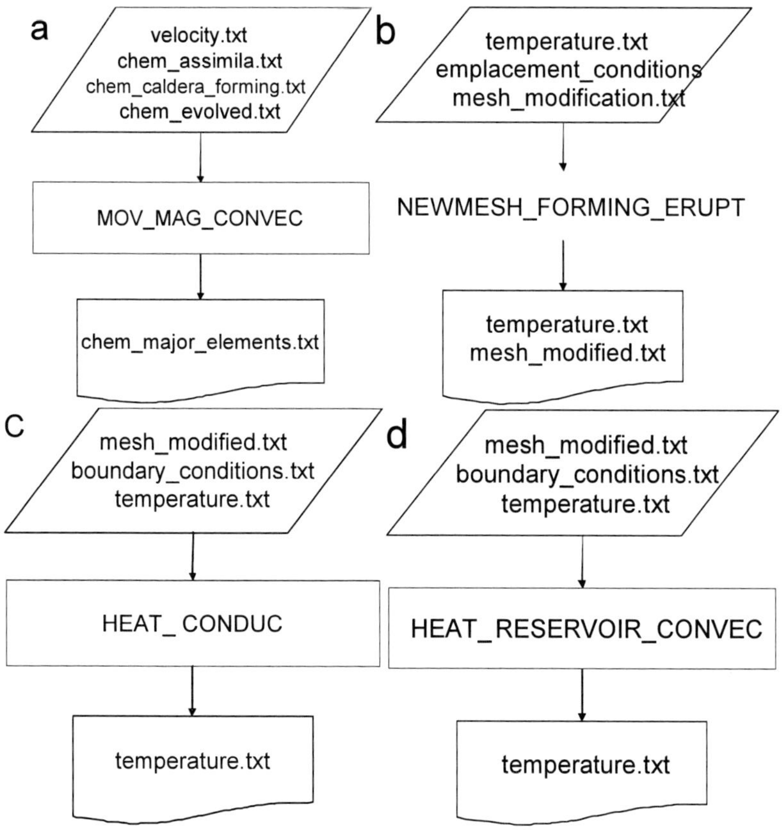

Figure 3. Explanation of the last four programs of TCHEMSYS computer software, with more explicit information on input and output files. (a) MOV_MAG_CONVEC; (b) NEWMESH_FORMING_ERUPT; (c) HEAT_CONDUC; and (d) HEAT_RESERVOIR_CONVEC.

A large number of studies report equations to model different magma chamber processes [Hansen and Yuen 1990; Valentine 1994; Spera et al. 1995; Folch and Martí 1998; Mourtada-Bonnefoi et al. 1999; Folch et al. 2001; Wallace and Bergantz 2002; Jellinek and DePaolo 2003]. Similarly, processes such as sedimentation and solidification in fluids other than magma [Koyaguchi et al. 1990; Ni and Beckermann 1991; Beckermann and Viskanta 1993; Ni and Incropera 1995a, b; Bergantz and Ni 1999], can help us better understand magma chamber processes. Melting of crustal rocks provides information on the assimilation processes of magma chambers [Huppert and Sparks 1988; Fountain et al. 1989; J. Marsh 1989; Bergantz 1992; Barboza and Bergantz 1997, 1998].

Details on probable magma chamber geometries and convective processes therein have been given by several researchers [Spera et al. 1982; Huppert and Sparks 1984; Campbell and Turner 1985; Lowell 1985; Brandeis and Jaupart 1986; Martin et al. 1987; Brandeis and Marsh 1989; Oldenburg et al. 1989, 1990; Rudman 1992; Bagdassarov and Fradkov 1993; Royer and Flores 1994; Bergantz 2000]. Replenishment of magma chambers either by denser [Snyder and Tait 1995] or lighter materials [Huppert et al. 1986; Weinberg and Leitch 1998] has also been studied to some extent. Additionally, an important role of volatile exsolution, movement, and control in heat transfer and eruption processes has been stressed [Huppert et al. 1982; Turner et al. 1983; Huppert and Woods 1988; Ida 1995; McLeod 1999; Simakin and Botcharnikov 2001; Mastin 2002; Bachman and Bergantz 2006]. The effects of viscosity in modeling of magma chambers have also been studied [e.g., Campbell and Turner 1986]. Similarly, effects of pressure and temperature on thermal conductivity have been reported by [Dubois et al. 1995].

Attempts to reproduce the volcanic history of Phlegraean fields in Italy through modeling of a magma chamber and the high temperatures measured in geothermal wells were presented more than 20 years ago [Giberti et al. 1984b]. On the other hand, computer programs, written in Fortran 77, for conductive cooling of dikes were also reported [Delaney 1988]. Seafloor black smokers were studied through heat transfer processes between a magma chamber and the overlying hydrothermal system [Lowell and Burnell 1991]. Most of these studies were carried out with the conductive process as the main heat dissipation mechanism, the convective process being sometimes simulated by increasing to many-fold the thermal diffusivity of the medium *without* actually simulating mass transport in the chamber or reservoir.

Double diffusive convection in a felsic magma chamber was modeled in two dimensions (2-D) [Bagdassarov and Fradkov 1993]. One dimensional (1-D) model of cooling and crystallization of sheet-like magma bodies was evaluated [Hort 1997]. A conductive model that takes into account the *in situ* heat contribution from natural radioactive elements (U, Th, and K) present in the geological medium, for the cooling of a magma chamber beneath the La Primavera caldera in Mexico was also used to predict shallow-level isotherms in this area [Verma and Rodríguez-González 1997]. A series of 2-D conductive/convective numerical models of possible magma chamber configurations were used to show that only a limited range of such configurations can predict the thermal regime at Campi Flegrei, Italy [Wohletz et al. 1999]. A computer program to model heat transfer processes in 2-D was also published [Bonneville and Capolsini 1999].

More recently, Clear Lake magmatic-hydrothermal system in California, USA has been thermally modeled in 2-D [Stimac et al. 2001]. These authors assumed a fairly elaborate and complex geological model for the entire crust and upper mantle beneath the Clear Lake area. This model consisted of a deep, dominantly-mafic system, and a shallow, dominantly-silicic system, with a mafic to felsic system lying at intermediate depths [see Fig. 4 in Stimac et al. 2001]. A 2-D mechanical-thermal fluid-dynamic model for geothermal systems at calderas, with an application to the Campi Flegrei, was also presented [Troise et al. 2001]. Excellent attempts to thermally constrain the chemical evolution of magma bodies have been reported [Bohrson and Spera, 2001, 2003; Spera and Bohrson 2001, 2002], although these authors did not provide any results of *in situ* modeling. Finally, a 2-D model of the magma chamber was recently used to simulate differentiation processes in a cooling magma body [Kuritani 2004]. Specific thermal modeling studies published for the Los Humeros caldera based on only the heat conduction process will be mentioned in the Application section of this chapter.

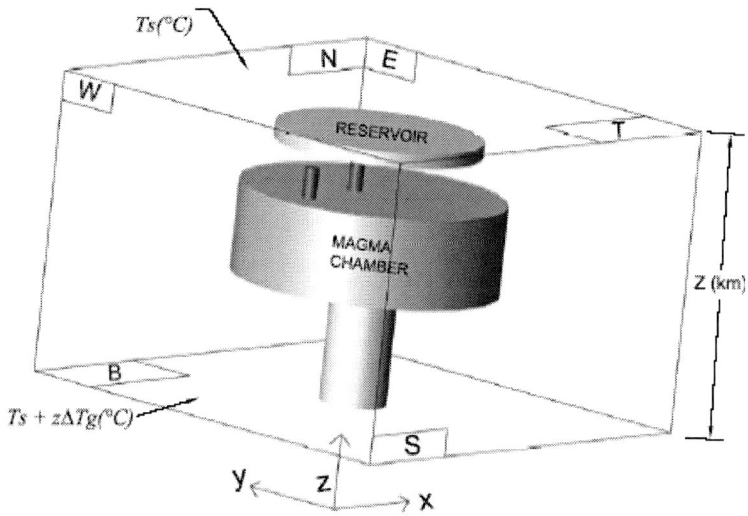

Figure 4. An schematic model in three-dimensions of the magma chamber and geothermal reservoir, emplaced in a larger region used for simulation in TCHEMSYS. E, W, N, S, T, and B refer to east, west, north, south, top, and bottom, respectively. X, Y, and Z represent the Cartesian coordinate system. The boundary conditions (temperature) for the top and bottom are, respectively, T_s and $T_s + z\Delta T_g$ (°C), where ΔT_g is the temperature gradient in °C/km and *z* is the total height in km of the simulated region.

Thus, the present study probably represents the first 3-D model to simultaneously simulate the thermal (both conductive and convective processes) and chemical aspects (processes such as crystallization, magma mixing, assimilation, recharge, and eruption) of a magma chamber in a geothermal system, and also to simulate convection in the overlying geothermal reservoir.

New Computer Program

To achieve an innovation in geothermal research, we have written a new computer program or software TCHEMSYS (thermal and chemical modeling of a volcanic-geothermal system), in Fortran 90, for the formulation and simulation of an integrated chamber-reservoir model of a geothermal field. This software simulates heat transfer conductive and convective processes in both magma chamber and geothermal reservoir and also computes *in-situ* major element chemistry of magmas that evolved in the magma chamber as a result of processes of assimilation, crystallization (liquid line of descent), magma mixing, recharge, and eruption. It is important to mention that this combined task is accomplished in three dimensions (3-D) – a substantial improvement as compared to the current practice of obtaining thermal solutions in

1-D or 2-D and of modeling chemical data obtained from the analysis of surface or drill well rocks *without* a clear reference to the actual location (within the Earth) where the magmas were stored prior to eruption. When temperature estimates in drill wells and chemical data for surface rocks are available from a geothermal field, this information can be used to constrain the model we are proposing as a new research frontier, as has been shown in this chapter for the Los Humeros geothermal field (see the Application section below).

The basic modular structure of the software is presented in Fig. 1. The entire program was successfully run, under the dynamic memory concept, on a personal computer (PC) with 0.98 Gb RAM, 2.40 GHz processor, and 80 Gb hard disk. At least similar computing facilities, especially the RAM size, will be actually required to successfully run TCHEMSYS on a different PC. We wish to stress that we did not purposely try a more sophisticated facility such as a work station, or several PCs linked to do a common job, because we wanted to ascertain that our 3-D software could be routinely used by us or by others interested in such an adventure, on a widely available facility, i. e., a simple PC.

This software is probably one of the first, if not the first, set of computer programs to do the combined job of thermal and chemical modeling in 3-D. More information on the modules of TCHEMSYS is schematically presented in Figs. 2a-d and 3a-d.

Description of Modules

(1) Heat_Forming

This program simulates in 3-D the thermal evolution of an area in which a magma chamber has been emplaced. The program first reads the initial conditions from three data files (boundary_conditions, emplacement_conditions, and mesh_construction, all in txt format; Fig. 2a). The mesh and simulation parameters are computed from these input data.

The boundary_conditions file includes the mean surface temperature (T_s °C) and the mean temperature gradient (ΔT_g °C/km) prior to the emplacement of a magma chamber. From these data, boundary conditions at all six boundaries of the simulated region (east, west, north, south, top, and bottom; Fig. 4) are inferred. For example, if the lower boundary (bottom) of the simulated region is at z km depth, its temperature will be fixed at ($T_s + z\Delta T_g$ °C; Fig. 4). Similarly, the vertical boundaries will be maintained at a certain temperature profile consistent with T_s °C and ΔT_g °C/km.

The data for the magma chamber, which is assumed to be of a cylindrical shape (Figs. 4 and 5), are read from the emplacement_conditions file. This file includes values for chamber parameters such as approximate volume (v_{cham}), radius (r_{cham}), depth of the top of the chamber (d_{cham}), and magma emplacement temperature (T_{cham}). The radius is assumed to be an integer-multiple of the mesh size, with the result that the chamber is represented by the same even number of control volumes in both x and y directions. The height of the cylinder (h_{cham}) is then calculated from the chamber volume (v_{cham}) and radius (r_{cham}), and is adjusted to fit an odd-number of control volume layers. Therefore, the chamber volume used in the modeling will only be approximately similar to v_{cham}. The magma feeding zone is also modeled as a cylinder with a radius r_{feed} and height h_{feed} [see e.g., Fountain et al. 1989; Oldenburg et al.

1989] and is schematically shown in Figs. 4 and 5. The dimensions of this zone are taken to be of a radius r_{feed} and height h_{feed}.

The total number of nodes in each of the three axes are determined from the input data on magma chamber dimensions, the mesh size, and the total volume used for thermal and chemical modeling (Figs. 4 and 5). The present version of the program is set for the number of mesh or control volumes of up to 120 in X-direction, 120 in Y-direction, and 80 in Z direction (with the resulting 1,152,000 control volumes for simulations!). These mesh numbers can be increased (and consequently the mesh size could be decreased) if a computer with a greater RAM were available for its use in simulations.

The mesh_construction file also provides information (size and physical properties) on up to 10 strata or geological layers overlying the magma chamber. More layers can be easily accommodated by modifying the input data in mesh_construction file. The thickness of each layer should bc the same as the vertical mesh size or an integer-multiple of this value, although layers of other thicknesses (different from the integer-multiples) could be accommodated by calculating and providing the combined physical property data for the different control volumes. Finally, values for time step Δt and total simulation time t are also read from this file. Note that the step Δt can have different values for the initial stage of magma chamber filling and for later stages of magma evolution. This flexibility is convenient to study the effects of the filling time parameter on the simulation results, by changing Δt for magma chamber-filling from a very fast to a fairly slow process, and at the same time, keeping the same modeling conditions for the later period.

The emplacement of the magma chamber is visualized as a growing body from the central layer to alternatively upper and lower layers. The thickness of each layer within the chamber is, in fact, the same as the mesh size in Z-direction. As stated earlier, for the central layer to be valid as presumed, an odd number of layers are assumed to schematically represent the chamber. During the process of magma chamber filling, no convection is assumed to take place; instead, only the conductive process accounts for heat transport.

The equations to be solved are presented in several books [Patankar 1980; Valentine 1992; Currie 1993; Versteeg and Malalasekera 1995]. The simplified equations for Cartesian coordinates $r=(x,y,z)$, velocity $\vec{u}=(u,v,w)$, are, respectively, for mass conservation (equation 1), momentum conservation (equations 2-4), and energy conservation (equation 5), as follows.

$$\frac{\partial u}{\partial x}+\frac{\partial v}{\partial y}+\frac{\partial w}{\partial z}=0 \tag{1}$$

$$\rho\left(\frac{\partial u}{\partial t}+u\frac{\partial u}{\partial x}+v\frac{\partial u}{\partial y}+w\frac{\partial u}{\partial z}\right)=-\frac{\partial p}{\partial x}+\mu\nabla^2 u+\rho f_x \tag{2}$$

$$\rho\left(\frac{\partial v}{\partial t}+u\frac{\partial v}{\partial x}+v\frac{\partial v}{\partial y}+w\frac{\partial v}{\partial z}\right)=-\frac{\partial p}{\partial y}+\mu\nabla^2 v+\rho f_y \tag{3}$$

$$\rho\left(\frac{\partial w}{\partial t}+u\frac{\partial w}{\partial x}+v\frac{\partial w}{\partial y}+w\frac{\partial w}{\partial z}\right)=-\frac{\partial p}{\partial z}+\mu\nabla^2 w+\rho f_z \tag{4}$$

$$\rho C_p\left(\frac{\partial T}{\partial t}+u\frac{\partial T}{\partial x}+v\frac{\partial T}{\partial y}+w\frac{\partial T}{\partial z}\right)=k\nabla^2 T \tag{5}$$

where the first and second terms on the right-hand side of equations (2-4) are surface forces, the third term represents the body forces, C_p is heat capacity at constant pressure, and

$$\nabla^2=\frac{\partial^2}{\partial x^2}+\frac{\partial^2}{\partial y^2}+\frac{\partial^2}{\partial z^2} \tag{6}$$

The following general equation represents the above equations for a property ϕ [Versteeg and Malalasekera 1995]:

$$\frac{\partial}{\partial t}(\rho\phi)+div(\rho u\phi)=div(\Gamma grad\phi)+S_\phi \tag{7}$$

Transient three-dimensional convection-diffusion of ϕ in a velocity field $\vec{u}=(u,v,w)$ is (where S is a source term):

$$\frac{\partial(\rho\phi)}{\partial t}+\frac{\partial(\rho u\phi)}{\partial x}+\frac{\partial(\rho v\phi)}{\partial y}+\frac{\partial(\rho w\phi)}{\partial z}=\frac{\partial}{\partial x}(\Gamma\frac{\partial\phi}{\partial x})+\frac{\partial}{\partial y}(\Gamma\frac{\partial\phi}{\partial y})+\frac{\partial}{\partial z}(\Gamma\frac{\partial\phi}{\partial z})+S \tag{8}$$

The discretization of this equation can be written as:

$$a_P\phi_P=a_W\phi_W+a_E\phi_E+a_S\phi_S+a_N\phi_N+a_B\phi_B+a_T\phi_T+a_P^0\phi_P^0+S_u \tag{9}$$

where

$$a_P=a_W+a_E+a_S+a_N+a_B+a_T+\frac{\rho_P^0\Delta V}{\Delta t}+\Delta F-S_P \tag{10}$$

$$S\Delta V=S_u+S_P\phi_P \tag{11}$$

The control volume method is used for solving these equations in three dimensions using line by line (in west to east direction) on a selected plane, and thus equation (9) can be simplified in terms of four coefficients a, b, c, and d as follows:

$$a\phi_P=b\phi_E+c\phi_W+d \tag{12}$$

This system of equations (12) for each control volume can be solved using TDMA [TriDiagonal-Matrix Algorithm; Patankar 1980].

After the chamber is completely filled, the HEAT_FORMING program calls for a subroutine veloz, which computes the velocity field of the magma chamber. This information is used to estimate the discretization parameters a, b, c and d. Note that all these parameters should be positive for a valid solution. Note also that the d parameter is highly sensitive to convection [Versteeg and Malalasekera 1995]. Two other coefficients p and q [Patankar 1980] were also computed before calculating the new temperature in each control volume. The 3-D problem was solved line by line on a selected plane and then the calculation was moved to the next plane, scanning the domain plane by plane. A very strict convergence test [see, for example, García-Valladares et al. 2006] was applied, in which, for our method, the temperatures calculated in two successive steps do not differ by more than 0.0000001°C. The program finally proceeds to write the results in two output data files, one for temperature and the other for velocity field (Fig. 2a).

The velocity field is solved in 3-D following the convection model in a vertical cylinder proposed by [Baïri 2003]. The solution of the velocity field inside the magma chamber is shown schematically in Fig. 5. Within a given ellipse the solution was obtained as done by several workers [e.g., Hansen and Yuen 1990; Oldenburg et al. 1990], whereas extrapolations and assumptions were used for control volumes lying outside the ellipse. We tested our velocity calculation subroutine by reproducing the velocity field data presented by [Baïri 2003, see Figs. 3-5 of this reference].

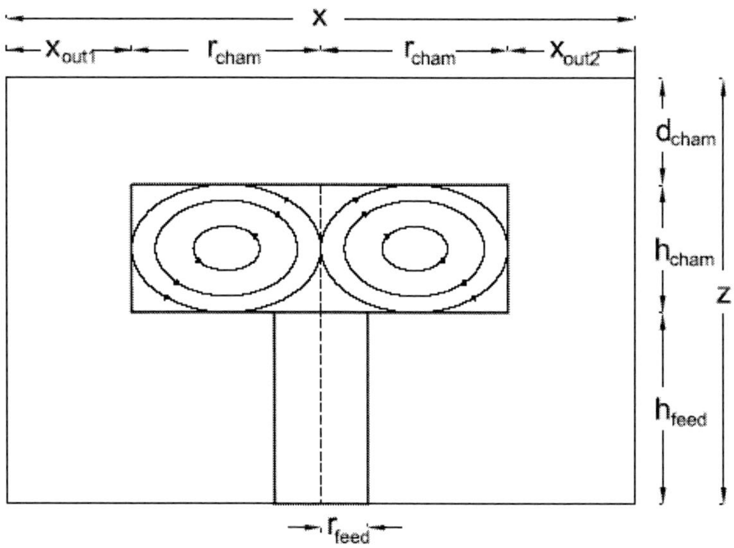

Figure 5. A simplified cross-section (vertical section at the center of the area) of the thermally and chemically modeled region, containing a cylindrical magma chamber with a radius r_{cham}, height h_{cham}, depth of the top of the magma chamber from the surface d_{cham}. The total region used for solving the thermal problem is supposed to have x, y, and z (km) in the X, Y, and Z directions, respectively. The feeding zone is also modeled as a cylinder with a radius r_{feed} and height h_{feed}. To obtain the total simulated length x in X direction, x_{out1} and x_{out2} are added to the total width ($2 \cdot r_{cham}$) of the magma chamber ($x=2 \cdot r_{cham}+x_{out1}+x_{out2}$), whereas the Y direction is established in a very similar manner. The vertical depth z of the modeled region is also shown as ($d_{cham}+h_{cham}+h_{feed}$).

(2) Balance_Mass_Forming

This program simulates the liquid line of descent resulting from crystallization of the magma in each control volume. Relevant petrological solutions have been given by several workers [e.g., Nielsen 1985, 1988a, b, 1992; Ghiorso 1985, 1991; Ghiorso and Carmichael 1985; Ariskin et al. 1993; Ghiorso and Sack 1995; Hirschmann et al. 1998]. We adopted Nielson's model [Nielsen 1988a, b] to process the most basic composition from the study area and to use the results of liquid line of descent for calculating mass balance in each control volume. If a basic magma such as a basalt is used as the starting magma composition in a magma chamber, Nielson's model is capable of handling several solid (mineral) phases; these are olivine, plagioclase, pyroxenes (orthopyroxene, augite and pigeonite), spinel, ilemnite, and apatite. From this petrological information the program computes the composition of all major elements in the remaining liquid as well as the amount (mass %) of all crystallized phases in a given time step (Δt) used for solving the geological problem. For all mineral groups the amounts of end-members were also estimated, for example, the fraction of forsterite and fayalite in olivine; of albite, anorthite, and orthoclase in plagioclase; and of enstatite, wollastonite, and ferrosilite in augite as well as in pigeonite.

This program reads input data from three txt files: temperature, physical_properties, and basic_crystallize. The physical_properties file includes information on molecular weights used in this module [Verma et al. 2003] and enthalpy data [Nicholls and Stout 1982; Richet and Bottinga 1984] as well as other physical properties required for a thermal and chemical solution of the model. These input data were used to estimate the heat contribution from the phases crystallized from the magma in each control volume, according to the following equation:

$$S_{ejmi} = (\rho_m \Delta x \Delta y \Delta z H_{ejmi} \mathrm{f}_{\mathrm{mi}} \mathrm{f}_{\mathrm{ejmi}})/\mathrm{MW}_{\mathrm{ejmi}} \tag{13}$$

where ρ_m is the density of magma, $\Delta x \Delta y \Delta z$ is the mesh volume, H_{ejmi} is enthalpy of melting of end-member j of mineral group i, f_{mi} is the mass fraction of mineral group i, $\mathrm{f}_{\mathrm{ejmi}}$ is the mass fraction of end-member j of mineral group i, and $\mathrm{MW}_{\mathrm{ejmi}}$ is the molecular weight of the same end-member j of mineral group i.

The mean composition of each mineral or mineral group (olivine, plagioclase, orthopyroxene, augite, pigeonite, spinel, ilmenite, and apatite) is calculated for a given time step Δt. All actions of this program are thus completed. The output data from this program are written in two txt files: chem_evolved and heat_source (Fig. 2b).

(3) Heat_Convec

This program simulates thermal evolution of a transient magma chamber in 3-D. Five input files (velocity, temperature, boundary_conditions, mesh_construction, and heat_source; Fig. 2c) are first read in this module. This program simulates both conductive and convective heat transfer processes and provides an output of resulting temperatures (temperature file; Fig. 2c) simulated in a given time step Δt or total time interval t. These two time parameters (Δt and t) can have values different from the heat_forming module.

Convection is assumed to commence after the chamber is filled up to a pre-established extent if the velocity regime within the chamber is appropriate for a convective process to take place. The filling of the chamber is assumed to be sufficiently fast so that convective processes can be safely neglected during this fill up process.

The sink subroutine of this program was written to take into account the heat required (consumed) for generating partial melts from the country rock, partial melting being an endothermic process.

(4) Balance_Mass_Convec

This program computes the mass balance in a given control volume. It reads from an input file basic_crystallize (output from Nielsen's liquid line of descent program), the % amount and major element chemical composition of all crystallizing solid phases and remaining liquid as well as temperature of each control volume (temperature file) and calculates the chemical and thermal parameters and writes them in chem_evolved and heat_source files (Fig. 2d).

The composition of most basic magma assumed to feed Nielsen's program [Nielsen 1992] is also read from the same basic_crystallize data file. The new chemical composition of residual magma in a given control volume is then computed from the input data. The % amounts (i.e., % mass) of end-members of different minerals crystallized in a given step are calculated in order to use these data to estimate heat contribution to the magmatic system in each control volume. The net (weighted) composition of crystallizing phases is also calculated for each control volume and for a given time step Δt.

After reading the data, the program calls heatsource subroutine to estimate the heat contribution from crystallizing phases in the magma chamber. This program takes into account the conditions for assimilation of the country rock, which if melted will also be available for eruption in addition to the evolved magma generated through crystallization mechanism. The heatsource subroutine also links the individual control volume temperatures to Nielsen's liquid line of descent output. This information is validated through certain tests such as the sum of all phases should be between 99.98 and 100.02, and if somewhat different from 100 (but within this range), the sum is adjusted to 100 to give new adjusted abundances of individual mineral phases. The chemical composition (consisting of 11 components) of the remaining magma is then computed. From the amounts of individual solid (crystalline) phases formed in a given time step Δt, the net heat contribution is computed for this exothermic process using an equation similar to eq. (13). This program gives two outputs in txt format: chem_evolved and heat_source.

(5) Mov_Mag_Convec

This program computes *in-situ* the major element composition in three dimensions (3-D) for each control volume by taking into account the velocity field in the magma chamber. It reads (Fig. 3a) velocity field from HEAT_FORMING program, chemical composition of the assimilant (chem_assimila) and of the caldera-forming main event (chem_caldera_forming), and output (chem_evolved) file from the earlier program BALANCE_MASS_CONVEC.

Because this program takes into account the movement of magma in 3-D, the mass contribution from a node situated at a corner of the magma chamber ($_{i-1,\,j-1,\,k+1}$) to the neighbor node ($_{i,\,j,\,k}$) is as follows:

$$S_{i-1,j-1,k+1} = fl_{i-1,j-1,k+1} \cdot \left|vx_{i-1,j-1,k+1}\right| \cdot \left|vy_{i-1,j-1,k+1}\right| \cdot \left|vz_{i-1,j-1,k+1}\right| \cdot \Delta t^3 \qquad (14)$$

where $fl_{(i-1,\ j-1,\ k+1)}$ is the fraction of liquid in the control volume $(_{i-1,\ j-1,\ k+1})$. Equation (14) holds provided the velocities vx and vy are positive and vz negative. This is simply one example out of the 27 such equations programmed in this software.

This program gives an output file in txt format: chem_major_elements.

(6) Newmesh_Forming_Erupt

This program takes into account the eruption of magma and modifies the magma chamber and surrounding area to generate new mesh conditions for further simulation. It reads (Fig. 3b) modified data from emplacement_conditions and mesh_modification files as well as temperature file from the last output of HEAT_CONVEC program (see Fig. 2c). The volcanic eruption was simulated by withdrawing the necessary amount of magma from the upper portion of the magma chamber and placing it at the surface of the modeled region. To simulate recharge of the magma chamber, some of the control volumes (in the magma chamber), from which magma for eruption was withdrawn, were assumed to be filled with new batch of basic magma, whereas the others not filled by recharge magma, account for the caldera collapse. The eruption process might also generate one or more conduits above the magma chamber, as will be shown in the Application section of this chapter.

The program gives a new temperature file as the output as well as new mesh (mesh_modified) information.

(7) Heat_Conduc

This program takes into account the heat conduction in the entire region, including the magma chamber and simulates the final temperature. It reads the output from the earlier program NEWMESH_FORMING_ERUPT (Fig. 3c). It must also read the boundary_conditions file. Because at a certain stage of magma chamber evolution, the magma might be fairly cooled and fragmented or unconnected due to the presence of abundant solid phases, this HEAT_CONDUC program is called for further simulations. However, if at any given stage the user decides that convection in the magma chamber was still taking place and hence should not be ignored, the HEAT_CONVEC program can be called to solve the problem instead of the HEAT_CONDUC program.

This HEAT_CONDUC program provides an output of the temperature regime as a txt file.

(8) Heat_Reservoir_Convec

This program takes into account convection in the geothermal reservoir and simulates the final thermal regime that can be interpreted to provide a series of isotherms. These isotherms or simulated temperatures can be compared with the thermal regime inferred from the static formation temperatures that are computed from the actually measured temperatures in boreholes using different algorithms [Luheshi 1983; Drury 1984; Shen and Beck 1986; Cao et al. 1988; Deming and Chapman 1988; Kutasov et al. 1988; Deming 1989; Hermanrud et al. 1990; Nielsen et al. 1990; Wilhelm 1990, 2000; Hasan and Kabir 1991, 1994; Prensky 1992; Santoyo et al. 2000; Kutasov 2003; Kutasov and Eppelbaum 2005; Fomin et al. 2003; Andaverde et al. 2005; Verma et al. 2006b, c].

The program reads the boundary_conditions, mesh_modified, and temperature files (Fig. 3d). The convection in the reservoir can be inferred from the depth-temperature profiles from the geothermal field, if available from independent estimates obtained in prior studies. Note that the thermal regime should preferably be obtained from static formation temperatures and not from the original, actually measured borehole temperatures. Convection in a geothermal reservoir can be simulated in exactly the same way as accomplished for the magma chamber (see HEAT_FORMING program above). The output from HEAT_RESERVOIR_CONVEC program contains information on the final temperatures in the entire modeled region.

In order to test the validity of thermal modeling from TCHEMSYS software, these simulated temperature results can be compared with the calculated static formation temperatures based on the extrapolation of actual borehole temperature measurements.

Application to the Los Humeros Geothermal Field, Puebla, Mexico

An application example of a Mexican geothermal field (Los Humeros, Puebla), currently under exploitation for electricity production, will highlight the use of our innovation strategy. This particular geothermal field was chosen for illustration purposes because of the availability of required thermal and chemical data to test the 3-D simulation model. We show in this chapter that we have successfully reproduced some of the chemical characteristics for the most voluminous caldera-forming eruption at about 0.46 Ma and the present-day thermal regime inferred from static formation temperatures using the quadratic regression methodology of the actually measured borehole temperature data proposed by [Andaverde et al. 2005]. We may also mention that this practice of "direct" modeling can be replaced in future by inverse modeling when greater computing and storage capacities of personal computers will be available.

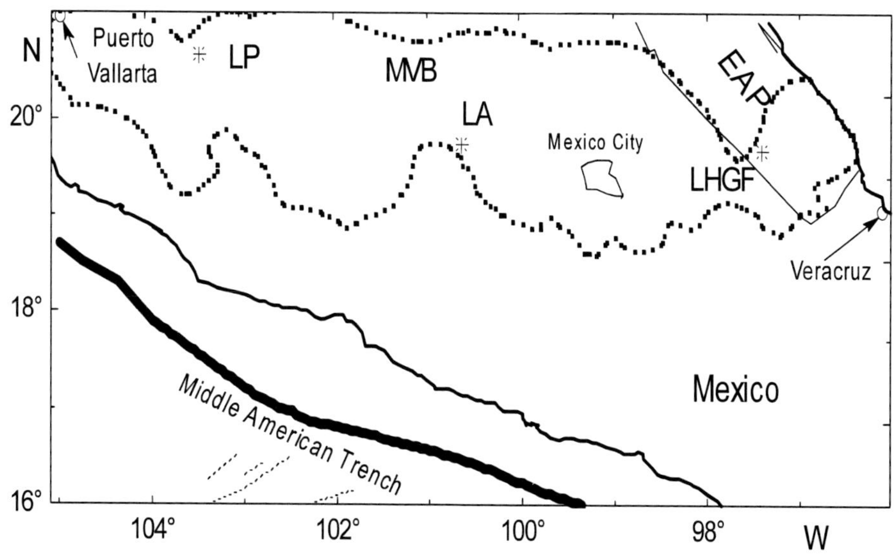

Figure 6. Location of the Los Humeros geothermal field (LHGF), Puebla, Mexico, in the Mexican Volcanic Belt (MVB). Two other geothermal fields, Los Azufres (LA) and La Primavera (LP), are also shown. EAP–Eastern Alkaline Province. Note that the LHGF is located on the overlapping region of the two volcanic provinces: MVB and EAP.

Los Humeros caldera (approximately 97°25'W longitude and 19°40'N latitude) is situated in the eastern part of the Mexican Volcanic Belt (Fig. 6). This caldera houses a geothermal field, which is the third largest electricity producing field in Mexico, with an installed capacity of about 35 MW [Hiriart and Andaluz 2000; Verma 2002a; internet site www.cfe.gob.mx, accessed on October 4, 2006]. The Los Humeros geothermal field (LHGF) has been the focus of a large number of studies in the field of: geology [Pérez Reynoso 1979; Ferriz 1985; Carrasco-Núñez and Branney 2005]; rock geochemistry [Verma and Lopez M. 1982; Verma 1983, 1984, 2000; Ferriz 1985; Ferriz and Mahood 1987]; geochronology [Ferriz and Mahood 1984]; geophysics [Flores L. et al. 1978; Mena and González-Morán 1978; Alvarez 1978; Ponce and Rodríguez 1978; González-Morán and Suro-Pérez 1982; López Mendiola and Munguía Bracamontes 1987; Campos-Enríquez and Arredondo-Fragoso 1992; Campos-Enríquez et al. 2005]; fluid geochemistry [Tello Hinojosa 1992a]; hydrothermal alteration [Arnold and Gonzalez Partida 1986, 1987; Arnold et al. 1986; Martínez Serrano and Alibert 1994; Bienkowski et al. 2005]; physical properties [Contreras et al. 1990]; well logging [Campos E. and Durán M. 1986; Medina Martínez 2000]; and reservoir studies [Ferriz 1982; Torres R. 1988, 1995; López Mendiola and Munguía B. 1989; Tello Hinojosa 1992b].

A preliminary attempt to conductively model the heat source beneath the LHGF, was made when only one drill well was available in this area [Prol and González-Morán 1982]. These authors assumed a much smaller heat source (a volume of about 100 km^3) than later inferred by [Verma 1985] from geochemical modeling of the volcanic activity from this area. Such a small magma chamber (100 km^3) will certainly not explain the voluminous caldera-forming eruption of Xáltipan ignimbrite and rhyolite flows (volume ~115 km^3) at about 0.46 Ma. Largely conductive thermal models were also obtained later for the LHGF assuming a much larger magma chamber [Verma et al. 1990; Castillo-Román et al. 1991; Verma and Andaverde 1995].

Crustal extension can largely facilitate the formation of magma chambers [Quick et al. 1994]. The extensional regime in the west-central to eastern part of the Mexican Volcanic Belt (MVB; Fig. 6) inferred from numerous geochemical studies [Verma 1994, 1999a, b, 2001, 2002b, 2003, 2004, 2006; Márquez et al. 1999, 2001; Velasco-Tapia and Verma 2001; Verma and Hasenaka 2004; Verma et al. 2006d] should, therefore, facilitate the emplacement of magma chambers in this area. The LHGF located in the eastern part of the MVB (Fig. 6) can thus be assumed to be underlain by a magma chamber. The modeling of rock geochemistry data [Verma 1985, 2000] also requires that the area be underlain by a large (~1500 km^3) magma body.

Input Data

An example of input data in a doc format is shown in Table 1, which presents the relevant data that are read from different txt files: boundary_conditions, emplacement_conditions, and mesh_construction. For example, for the LHGF initially the surface temperature was assumed to be 25°C, with an initial temperature gradient of 30°C/km (Table 1). Other data needed in the TCHEMSYS software are molecular weights of major oxides (Table 2) and thermophysical properties such as enthalpy of melting of crystallizing minerals (Table 3).

Table 1. Boundary and initial conditions used for Los Humeros geothermal field (LHGF) modeling

Physical property (units)	Emplacement (at 0.486 Ma)	Convection in magma chamber (between 0.485 and 0.460 Ma)	Conduction in magma chamber (between 0.460 and the present)	Convection in geothermal reservoir (between 0.020 Ma and the present)
boundary_conditions				
surface temperature (T_s °C)	25	25	25	25
temperature gradient (ΔT_g °C/km)	30	30	30	30
emplacement_conditions				
volume (v_{cham}) (km^3)	1500	1500	1500	1500
radius (r_{cham}) (m)	8500	9000	9000	9000
depth of the top of the chamber (d_{cham}) (m)	4750	5000	5000	5000
magma emplacement temperature (T_{cham})(°C)	1350	1350	1350 (1100 in conduits)	---
Initial velocity in the magma chamber (m/s)		$1x10^{-9}$		
mesh_construction				
Length - x (m)	30000	30000	30000	30000
Number of control volumes in X-direction	120	120	120	120
Length - y (m)	30000	30000	30000	30000
Number of control volumes in Y-direction	120	120	120	120
Number of geological strata	4	4	4	5
Delta-z (δz) (m)	250	250	250	250
Strata 1 width (m)	17000	17000	17000	17000
Thermal conductivity (W/m K)	2.853	2.853	2.853	2.853
Specific heat (J/kg K)	914	914	914	914
Density (kg/m^3)	2680	2680	2680	2680
Strata 2 width (m)	1000	1000	1000	1000
Thermal conductivity (W/m K)	2.705	2.705	2.705	2.705
Specific heat (J/kg K)	854	854	854	854
Density (kg/m^3)	2180	2180	2180	2180

Table 1. Continued

Physical property (units)	Emplacement (at 0.486 Ma)	Convection in magma chamber (between 0.485 and 0.460 Ma)	Conduction in magma chamber (between 0.460 and the present)	Convection in geothermal reservoir (between 0.020 Ma and the present)
Strata 3 width (m)	1000	1000	1000	1000
Thermal conductivity (W/m K)	1.673	1.673	1.673	1.673
Specific heat (J/kg K)	1009	1009	1009	1009
Density (kg/m^3)	2394	2394	2394	2394
Strata 4 width (m)	1000	1000	1000	1000
Thermal conductivity (W/m K)	1.738	1.738	1.738	1.738
Specific heat (J/kg K)	885	885	885	885
Density (kg/m^3)	2360	2360	2360	2360
Time step Δt (year)	40	1,000	1,000	1,000
Total simulation time t (year) *	1,000	25,000	460,000 *	20,000

* These times are those for which a given process was reported in this chapter. The total simulated time was 486,000 years that reproduces the entire volcanic history of the LHGF. Note the conduction in the magma chamber and the convection in the geothermal reservoir were simultaneously operative during the last 20,000 years.

These thermophysical properties were taken either from direct measurements on drill core samples from the LHGF [Contreras et al. 1990] or from the literature when actual measured values were not available [e.g., Nicholls and Stout 1982; Kojitani and Akaogi 1997; Spera 2000; Akaogi et al. 2004; see also Table 3 for more details].

Table 2. Molecular weight data for major element oxides used in modeling [taken from Verma et al. 2003]

Oxide symbol	Oxide molecular weight (AMU)
SiO_2	60.0843
TiO_2	79.8658
Al_2O_3	101.961276
Fe_2O_3	159.6882
FeO	71.8444
MnO	70.937449
MgO	40.3044
CaO	56.0774
Na_2O	61.97894
K_2O	94.1960
P_2O_5	141.944522

The magma chamber dimensions are shown in Fig. 7 (refer to the km scale in the X-Z croos-section; the same scale applies to all directions). The mesh size used in these simulations were 0.25 km in each X, Y, or Z direction. Fig. 7 also gives other details of the entire region thermally and chemically modeled in this work.

Table 3. Enthalpy of melting data for crystallizing minerals

Mineral	Enthalpy of melting (J/gfw)
Fayalite [a]	92,174
Forterite [a]	170,193
Albita [a]	65,354
Anorthite [a]	135,562
Orthoclase [a]	58,492
Ilmenite [a]	90,667
Wollastonite [b]	86,520
Enstatite [c]	75,300
Ferroselite [d]	50,000
Spinel [d]	50,000
Apatite [d]	50,000

[a] [Nicholls and Stout 1982]
[b] [Akaogi et al. 2004]
[c] [Kojitani and Akaogi 1997]
[d] Assumed (some values had to be assumed).

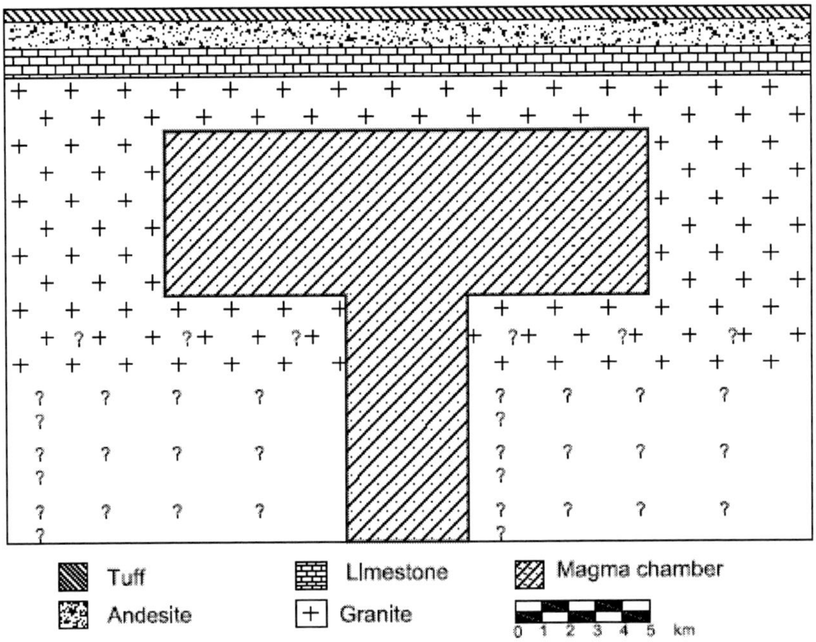

Figure 7. A simplified cross-section (vertical section at the center of the area) of the thermally and chemically modeled region of the Los Humeros geothermal field (LHGF). The dimensions and other parameters used for simulations with the TCHEMSYS software can be inferred from the scale which is same in all directions. This schematic diagram represents the LHGF after the magma chamber filling. Geological units are also schematically shown.

Table 4. A very small part of the output of Nielsen's liquid line of descent program for an initial magma composition of the most basic magma (HF117) from the Los Humeros geothermal field (LHGF)

Major oxide (%m/m)	Magma temperature (°C)									
	1320	1314	1220	1136	1135	1092	1075	1074	1046	1027
SiO_2	49.45	49.52	50.25	51.81	51.80	51.74	58.81	59.31	65.25	66.78
TiO_2	1.23	1.23	1.33	2.33	2.34	3.14	1.59	1.50	0.70	0.55
Al_2O_3	16.14	16.24	17.50	13.95	13.95	13.12	13.91	14.01	14.81	14.92
Fe_2O_3	1.51	0.95	1.05	1.82	1.83	2.36	1.77	1.71	0.90	0.66
FeO	7.56	8.07	7.80	9.79	9.85	11.84	6.89	6.55	2.98	2.10
MnO	0.02	0.02	0.02	0.03	0.03	0.03	0.02	0.02	0.02	0.01
MgO	11.17	10.97	8.02	4.69	4.65	2.98	1.35	1.25	0.33	0.10
CaO	8.99	9.05	9.75	10.19	10.13	8.18	6.67	6.53	4.86	4.59
Na_2O	3.26	3.28	3.54	4.20	4.22	4.91	6.29	6.38	7.25	7.30
K_2O	0.49	0.49	0.54	0.84	0.85	1.17	1.80	1.84	2.38	2.59
P_2O_5	0.18	0.18	0.20	0.35	0.35	0.53	0.90	0.90	0.52	0.40
Sum	100	100	100	100	100	100	100	100	100	100
Total % crystallized	0	0.6	9.1	49.6	50.1	67.1	80.5	81.0	86.2	87.9

Table 4. Continued

Major oxide (%m/m)	Magma temperature (°C)									
	1320	1314	1220	1136	1135	1092	1075	1074	1046	1027
% ol cryst.	0	0.5	0.2	0.1	0	0	0	0	0	0
% opx cryst.	0	0	0	0	0	0	0	0	0	0.1
% aug cryst.	0	0	0	0.1	0.3	0.2	0	0.2	0	0
% plg cryst.	0	0	0.3	0.3	0.2	0.2	0.3	0.2	0.4	0.4
% sp cryst.	0	0	0	0	0	0.1	0.1	0.1	0.06	0.04
% pig cryst.	0	0	0	0	0	0	0.1	0	0.1	0
% ap cryst.	0	0	0	0	0	0	0	0.01	0.02	0.01

Total % crystallized–total amount of minerals formed upto a given temperature (see the temperature value in the column heading); ol–olivine; opx–orthopyroxene; aug–augite; plg–plagioclase; sp–spinel; pig–pigeonite; ap–apatite; cryst.–mineral crystallized.

The major element composition of the most basic magma HF117, analyzed by [Verma and López M. 1982; Verma 1983, 1984, 2000], erupted in this area as the latest event was processed using Nielsen´s algorithm and computer program [Nielsen 1985, 1988a, b, 1992] to obtain the temperature-dependent amount (mass %) and major element composition of the crystallizing phases as well as the remaining liquid. However, the initial composition of HF117 was first adjusted to 100% on an anhydrous basis after dividing total Fe into both oxidation varieties using the SINCLAS computer program [Verma et al. 2002]. This initial composition after adjustment is given in the first data column under the heading of 1320°C magma temperature (Table 4). The remaining columns give the chemical composition of the residual magma at a given temperature as well as the % mass of minerals crystallizing at this temperature step. Note that the chosen temperatures tabulated (Table 4) are for the steps with new minerals entering into the crystallization mechanism and older ones disappearing.

Nielsen's output file was too large to be practicable to reproduce here; it provided outputs for very small temperature steps for the cooling of a basic magma. However, Nielsen's liquid line of descent output for the basic magma HF117 from the LHVF generated some discrepancies at the final stages of magma evolution, with Mg concentration of the remaining magma being negative. Some minor corrections below the temperature of 1055°C in the amount of minerals crystallizing were assumed, according to which minor amounts of crystallizing olivine were assigned to crystallizing plagioclase. The Nielsen's output file was modified to include these minor corrections incorporated for the final few stages of magma evolution, which was actually fed to the TCHEMSYS software. Nielsen's file also contains compositions of crystallizing phases, which are not included in the example of Table 4.

For simulating the assimilation and eruption processes, chemical composition data (Table 5) are read. The simulation results presented below are divided into two parts: chemical and thermal.

Table 5. Composition of an assimilant and statistical parameters for the caldera-forming Xaltipán ignimbrite eruption in the Los Humeros geothermal field (LHGF)

Major Oxide	Assimilant Granite [a]	Ignimbrita Xaltipán [b] n	$\bar{x} \pm s$	99%CL	R
SiO_2	68.16	9	70 ± 5	64 - 76	60 - 76
TiO_2	0.60	9	0.47 ± 0.40	0.03 - 0.92	0.05 - 1.13
Al_2O_3	15.07	9	15.5 ± 1.5	13.3 - 16.8	12.4 - 16.6
Fe_2O_3	3.48	9	3.0 ± 1.9	0.9 - 5.1	1.2 - 6.2
MnO	0.088	9	0.052 ± 0.031	0.020- 0.092	0.020- 0.100
MgO	1.27	9	0.9 ± 1.1	0.0 - 2.1	0.1 - 3.4
CaO	1.38	9	1.9 ± 1.7	0.0 - 3.9	0.4 - 5.2
Na_2O	4.89	9	3.8 ± 0.5	3.3 - 4.4	3.1 - 4.5
K_2O	3.64	9	4.6 ± 1.1	3.3 - 5.8	2.8 - 5.8
P_2O_5	0.148	9	0.13 ± 0.11	0.01 - 0.28	0.05 - 0.36

Original data taken from the following sources: [a] [Verma 2000]; [b] [Ferriz 1985; Ferriz and Mahood 1987]. n–number of analyses; $\bar{x}$ –mean; s –standard deviation; 99%CL–confidence interval of the mean at 99% confidence level; R – range of all values [for more details see Verma 2005].

Simulation Results

(1) Chemical Modeling

The functioning of the TCHEMSYS computer software was tested by its application in the LHGF. The most voluminous caldera-forming eruption, Xáltipan ignimbrite, about 115 km^3 [Ferriz and Mahood 1984, 1987; Ferriz 1985], at 0.46 Ma, was used for this purpose. Statistical information on major element chemical data actually measured in ignimbrite samples from this eruption are synthesized in Table 5.

We used the TCHEMSYS software to carry out simulations after the emplacement of the magma chamber in the LHGF assumed to have occurred at about 0.486 Ma, about 0.026 m.y. before the voluminous Xáltipan ignimbrite eruption about 0.46 Ma. Figs. 7 and 8 show the simplified combined "magma chamber-reservoir" model for before and after Xáltipan ignimbrite eruption, respectively. The initial composition of the magma filling the magma chamber was assumed to be that of the most basic basaltic magma (HF117; Table 4), which actually erupted in the LHGF at <0.02 Ma. In order to check if the magma chamber and the surrounding rocks could evolve to generate magma compositions broadly similar to the Xáltipan ignimbrite, we identified the control volumes that had evolved to compositions lying within the 99%CL (Table 5) for this ignimbrite. The results for four selected major elements or oxides (SiO_2, Al_2O_3, MgO, and CaO) are shown in Fig. 9. Evolved magmas in a large number of control volumes achieve Xáltipan ignimbrite compositions in about 0.025 m.y. after chamber emplacement. Most other major elements showed a similar behavior although for some of them a considerably lesser number of control volumes having Xáltipan ignimbrite compositions were observed in our simulations. To obtain better agreement for all elements in the space parameters and to achieve the ignimbrite to have erupted as a single event, we may have to change some of the input parameters that could increase the number of control volumes having Xáltipan ignimbrite compositions (Fig. 8). We note that this chemical modeling is simply a first step to show the feasibility of such simulations, which can obviously be refined in future to obtain a better agreement with the chemistry of erupted products.

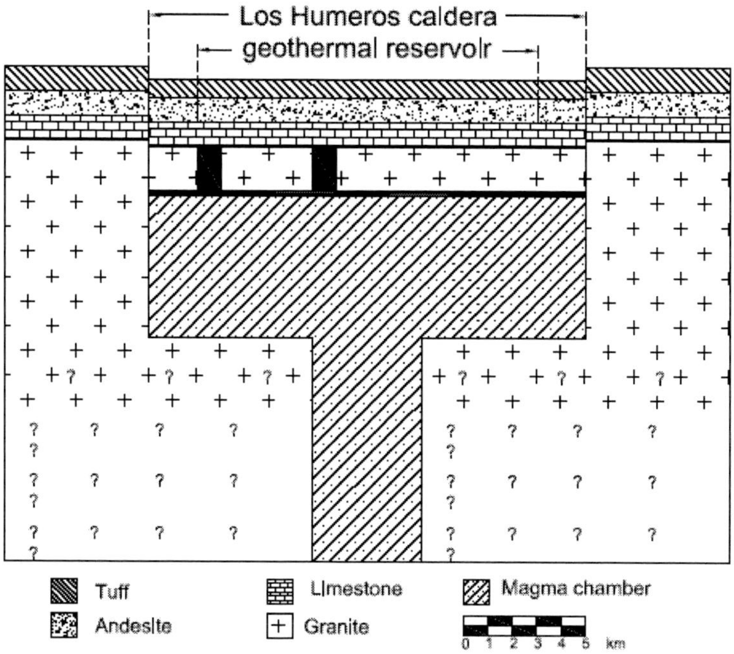

Figure 8. A simplified cross-section (vertical section at the center of the area) of the thermally and chemically modeled region of the Los Humeros geothermal field (LHGF). The dimensions and other parameters used for simulations with the TCHEMSYS software can be inferred from the scale which is same in all directions. This schematic diagram represents the LHGF after the eruption of caldera-forming Xáltipan ignimbrite at 0.46 Ma, which caused the formation of Los Humeros caldera (see the caldera collapse shown schematically in the upper part of the diagram). The two conduits inferred from geophysics [Flores L. et al. 1979; Mena and González-Morán 1979; Verma et al. 1990] are also integrated in the simulation model. The extent of the geothermal reservoir as inferred from well drilling is also shown. Compare and contrast with Fig. 7.

Compositional zonation in silicic magma chambers beneath calderas has been frequently proposed by several workers [e.g., Spera and Crisp 1981; Mahood 1984; Ferriz and Mahood 1984, Ferriz 1985; Grunder and Mahood 1988; Stix and Gorton 1993]. From the study of compositionally zoned eruptions [Trial and Spera 1990] inferred that evolved magma is generated at a rate between 10^{-2} and 10^{-4} km^3/a, with a typical value of about 10^{-3} km^3/a. These authors also considered that melting of silicic country rocks by intrusion of basaltic magma is a viable hypothesis because evolved melt is produced at a rate consistent with the volcanological observations. This means that to produce Xaltipan ignimbrite type caldera-forming event of about 115 km^3 of evolved magma we need a time period from 11,500 years to as much as 1.15 m.y. The results obtained in this chapter (partly presented in Fig. 9) show that the maximum number of nodes with the desired Xaltipan-type compositions reached in about 25,000 years (0.025 m.y.) after chamber emplacement, envisioned to have taken place in about 1000 years (0.001 m.y.). The geographical distribution of the control volumes with Xaltipan-type compositions partly supports this stratification model although more detailed simulation work is needed to fully evaluate its validity in the LHGF.

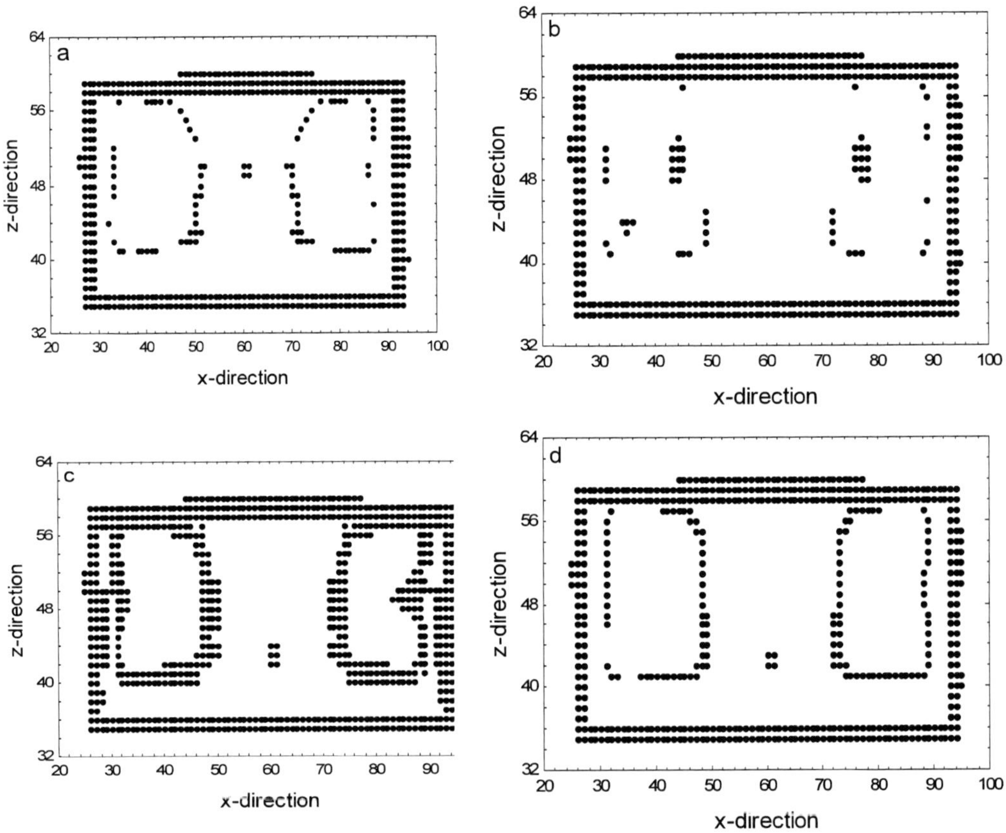

Figure 9. Chemical compositions of the LHGF magma chamber inferred after 25,000 years (0.025 m.y.) of simulation after magma chamber emplacement during 1,000 years. The nodes or control volumes that satisfy a certain chemical similarity condition are shown using filled circles. This condition is that the value of a given chemical parameter for the residual magma is similar to the chemical data for the Xáltipan ignimbrite (Table 5). In other words, if a given major element or oxide, e.g., SiO_2, has a concentration in the residual magma that lies within the 99%CL for that particular element or oxide (Table 5), that particular control volume is identified as the one suitable to provide magma for this voluminous eruption. (a) SiO_2; (b) Al_2O_3; (c) MgO; and (4) CaO.

In the present simulations, the crystallized solids and residual magmas were assumed to move at the same velocity within the chamber. This assumption can be refined in future to move these two phases at different velocities, and this distinction could be made even among the different minerals, which, depending on the type and size, could move at different velocities. Thus, "lighter" (low density) minerals could preferentially move towards the roof of the magma chamber, whereas the "heavier" minerals would concentrate towards the chamber floor. A much greater computing facility, however, will be required for this purpose.

Combined assimilation and fractional crystallization (AFC) processes might control the liquid line of descent [Kaneko and Koyaguch 2004] and should, therefore, be considered in future simulations. Magma chambers, particularly those of basaltic composition such as the one envisioned for the LHGF, might be replenished by an influx of magma whose density is less than that of the residual magma [Huppert et al. 1986]. This might be the case, for example, of the molten granitic rocks in our study area whose incorporation into the magma chamber should be considered in future both from thermal and chemical points of view. On the other hand, the injection process involving replenishment of a magma chamber with a denser liquid should also be specifically treated [Snyder and Tait 1995].

In this chapter, although we have considered only the major elements in our chemical modeling, all other chemical elements, especially the so-called trace elements and "volatile" elements, as well as petrogenetically useful isotopic ratios such as $^{87}Sr/^{86}Sr$, $^{143}Nd/^{144}Nd$, $^{206}Pb/^{204}Pb$, $^{207}Pb/^{204}Pb$, and $^{208}Pb/^{204}Pb$, should be taken into account in future in order to obtain a more complete picture of *in-situ* geochemical processes. This will make the modeling exercise really worthwhile to open a new direction in geochemical research.

(2) Thermal Modeling

The events of most voluminous Xáltipan ignimbrite eruption accompanied by a less voluminous recharge of basic magma in the magma chamber and the consequent caldera formation were simulated schematically to assume a new mesh construction as presented in Fig. 9. This event was assumed to have occurred at about 0.46 Ma based on actual geochronology data [Ferriz and Mahood 1984, 1987]. The TCHEMSYS software was used to carry out thermal simulation of magma chamber for about 0.44 m.y. (from 0.46 Ma to 0.02 Ma). During this period, only conductive processes were assumed to take place in the entire medium, i.e., both in the magma chamber and the geothermal reservoir. Then, at about 0.02 Ma, concurrent with the eruption of most basic basaltic magma being the last eruptive event to have taken place in the LHGF [Ferriz and Mahood 1984, 1987], the onset of convection in the geothermal reservoir was assumed to have initiated. Information of the exact timing of this later event is still lacking in the LHGF although methodology is available to achieve this in future through proper studies [Del Moro et al. 1982; Sturchio and Binz 1988; Sturchio et al. 1992; Lalou et al. 1993; Rihs et al. 2000]. From the simulated thermal data to the present day, i.e., as a result of simulation from 0.0486 Ma to 0 Ma, 200°C – 600°C isotherms were constructed (Fig. 10a). The "locally" different temperatures (see Fig. 10b) resulting from the influence of conduits are lost (smoothed out) when a very large region such as that of Fig. 8 is involved in such isotherm constructions using conventional computer programs.

To highlight the effects of magma conduits in the LHGF, we present in Fig. 10b the simulated temperatures at a depth of about 2000 m (below surface), in which the conduit influence on the thermal regime is clearly observed. To validate the above simulation results, the static formation temperatures (SFT) for the LHGF were obtained using the methodology of quadratic regressions recently proposed by [Andaverde et al. 2005]. The borehole temperatures actually measured in the LHGF were first compiled from several sources [Prol and González-Morán 1982; Campos E. and Durán M. 1986; Verma et al. 1990; Castillo-Román et al. 1991; Martínez Serrano and Alibert 1994; Medina Martínez 2000]. From these, SFT values were estimated and a synthesis of these data for 2000 m is also presented in Fig. 10b to facilitate a visual comparison with the simulated and static formation temperatures. A general agreement between the simulated and actually measured thermal regimes can be inferred at least for 2000 m depth. In future, these two sets of data can be better evaluated

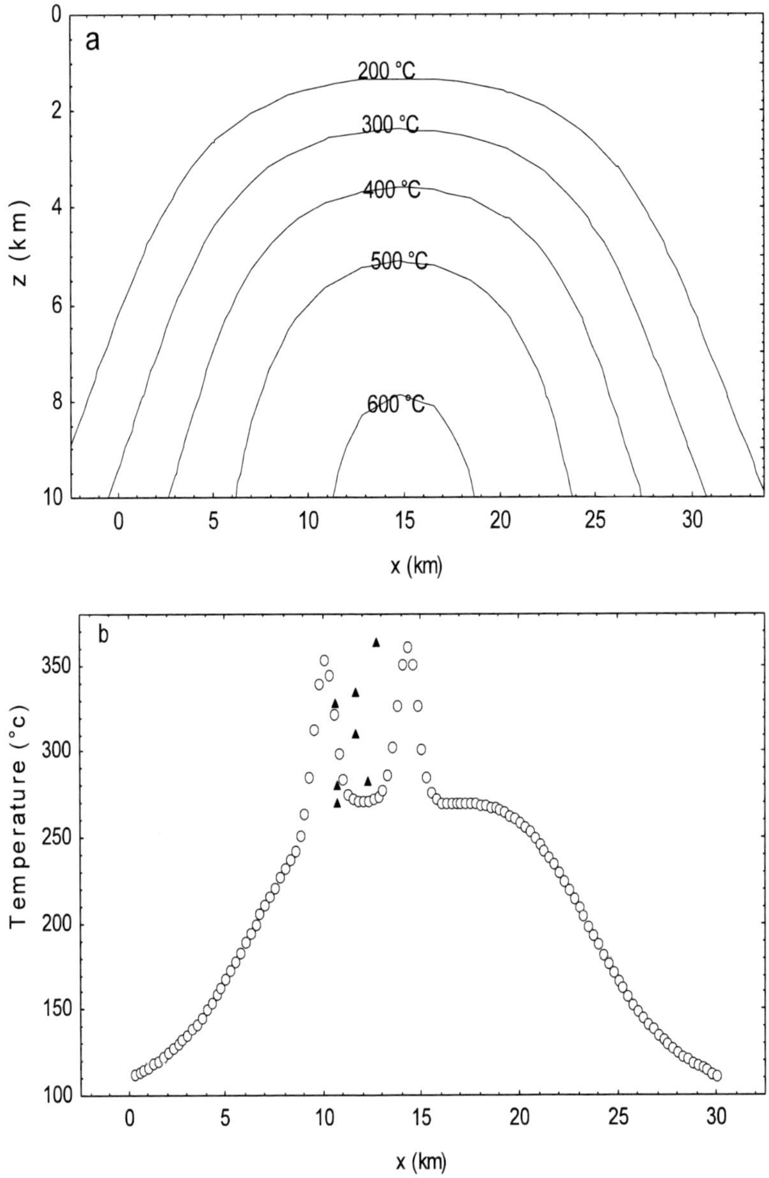

Figure 10. Final simulated results of thermal simulation in the Los Humeros geothermal field (LHGF) after about 0.486 m.y. of magma chamber evolution, taking into account the entire volcanic history of this area. The processes of assimilation, crystallization, convection, eruption, and recharge were assumed to be dominant during 0.485 Ma to 0.46 Ma, as well as convection in the geothermal reservoir for about 0.02 Ma. Both the most voluminous Xáltipan ignimbrite caldera-forming event at about 0.46 Ma and the latest basic magma eruption event at about 0.02 Ma have been simulated. (a) Present-day isotherms (200°C–600°C) simulated for the entire region during 0.486 m.y. of evolution; (b) An expanded view of the region to show simulated temperature regime at a subsurface depth of 2000 m (using open circles) to highlight the influence of the two conduits in the shallower part of the modeled region, see for comparison the actual static formation temperature estimates from borehole temperature measurements shown as filled triangles.

through proper statistical techniques [Verma 2005]. Furthermore, a "misfit" parameter can be calculated and used to refine the simulation data through inversion techniques. In fact, the timing of the final eruption event and the onset of convection, both poorly constrained in the LHGF would exert a large influence on the final thermal regime simulated by the TCHEMSYS software.

Finally, it is obvious that more programming as well as process-oriented work is needed before this line of research could provide more reliable applications and actual economic benefits during geothermal energy exploration and exploitation stages.

Conclusion

We conclude that the 3-D computer program for combined thermal and chemical modeling is an easy to use tool, which should find an important place in geothermal exploration and exploitation research. The usefulness of this new research frontier is illustrated from the application of our new computer program to the Los Humeros geothermal field, Puebla, Mexico.

Acknowledgments

We are most indebted to Alfredo Quiroz Ruiz for constantly helping us with useful tips to run our computer program for 3-D integrated model with the presently available limited capacity of personal computers. Octavio García Valladares offered help and guidance to the second author (JA) during the program writing and validation stages of our work, for which we are most grateful to him. During the final stage of this work JA received a Ph.D. scholarship from DGAPA-PAPIIT project IN104703-3 to Edgar Santoyo and Kailasa Pandarinath. Finally, Mirna Guevara is thanked for help with maintenance of our Geochem.lib database, this being an important feature to achieve an efficient handling of references listed in this chapter.

References

Akaogi, M., Yano, M., Tejima, Y., Iijima, M. & Kojitani, H. (2004). High-pressure transitions of diopside and wollastonite: phase equilibria and thermochemistry of $CaMgSi_2O_6$, $CaSiO_3$ and $CaSi_2O_5$-$CaTiSiO_5$ system. *Physics of the Earth and Planetary Interiors,* 143-144, 145-156.

Álvarez, R. (1978). Telluric, self-potential, and surface temperature profiles on Los Humeros caldera. *Geofisica Internacional,* **17**, 445-460.

Andaverde, J., Verma, S. P. & Santoyo, E. (2005). Uncertainty estimates of static formation temperatures in boreholes and evaluation of regression models. *Geophysical Journal International,* **160**, 1112-1122.

Andaverde, J., Verma, S. P. & Shildknecht, F. (1993). Aporte de calor por cristalización fraccionada en dos campos geotérmicos del Cinturón Volcánico Mexicano. *Geofisica Internacional,* **32**, 331-339.

Anguita, F., Verma, S. P., Márquez, A., Vasconcelos-F., M., López, I. & Laurrieta, A. (2001). Circular features in the Trans-Mexican Volcanic Belt. *Journal of Volcanology and Geothermal Research,* **107**, 265-274.

Ariskin, A. A., Frenkel, M. Y., Barmina, G. S. & Nielsen, R. L. (1993). COMAGMAT: a fortran program to model magma differentiation processes. *Computers & Geosciences,* **19**, 1155-1170.

Arnold, M. & Gonzalez Partida, E. (1986). Capacités oxydantes d'un fluide hydrothermal en ébullition. *Comptes Rendus de L Academie des Sciences du Paris,* **303**, 817-820.

Arnold, M. & Gonzalez Partida, E. (1987). Le système hydrothermal actuel de Los Humeros (Mexique): etat du système SO_4^{--} — SH_2 à 300°C, origine du soufre et phénomènes d'oxydation associés à l'ébullition du fluide ascendant. *Minerallium Deposita,* **22**, 90-98.

Arnold, M., Gonzalez Partida, E. & Munguia Bracamontes, F. (1986). Le système sulfate-sulfure: preuves d'un équilibre chimique et isotopique à 300°C; conséquences géochimiques. *Comptes Rendus de L Academie des Sciences du Paris,* **303**, 1097-1099.

Bachmann, O. & Bergantz, G. W. (2006). Gas percolation in upper-crustal silicic crystal mushes as a mechanism for upward heat advection and rejuvenation of near-solidus magma bodies. *Journal of Volcanology and Geothermal Research,* **149**, 85-102.

Bagdassarov, N. S. & Fradkov, A. S. (1993). Evolution of double diffusion convection in a felsic magma chamber. *Journal of Volcanology and Geothermal Research,* **54**, 291-308.

Baïri, A. (2003). Transient natural 2D convection in a cylindrical cavity with the upper face cooled by thermoelectric Peltier effect following an exponential law. *Applied Thermal Engineering,* **23**, 431-447.

Barbier, E. (1985). Geothermal energy in the context of energy in general and electric power supply. National and international aspects. *Geothermics,* **14**, 131-141.

Barboza, S. A. & Bergantz, G. W. (1997). Melt productivity and rheology: complementary influences on the progress of melting. *Numerical Heat Transfer, Part A,* **31**, 375-392.

Barboza, S. A. & Bergantz, G. W. (1998). Rheological transitions and the progress of melting of crustal rocks. *Earth and Planetary Science Letters,* **158**, 19-29.

Battaglia, M., Segall, P., Murray, J., Cervetti, P. & Langbein, J. (2003a). The mechanics of unrest at Long Valley caldera, California: 1. Modeling the geometry of the source using GPS, leveling and two-color EDM data. *Journal of Volcanology and Geothermal Research,* **127**, 195-217.

Battaglia, M., Segall, P. & Roberts, C. (2003b). The mechanics of unrest at Long Valley caldera, California: 2. Constraining the nature of the source using geodetic and micro-gravity data. *Journal of Volcanology and Geothermal Research,* **127**, 219-245.

Beckermann, C. & Viskanta, R. (1993). Mathematical modeling of transport phenomena during alloy solidification. *Applied Mechanics Review,* **46**, 1-27.

Bergantz, G. W. (1992). Conjugate solidification and melting in multicomponent open and closed systems. *International Journal of Heat and Mass Transfer,* **35**, 533-543.

Bergantz, G. W. (1995). Changing techniques and paradigms for the evaluation of magmatic processes. *Journal of Geophysical Research,* **100**, 17603-17613.

Bergantz, G. W. (2000). On the dynamics of magma mixing by reintrusion: implications for pluton assembly processes. *Journal of Structural Geology,* **22**, 1297-1309.

Bergantz, G. W. & Ni, J. (1999). A numerical study of sedimentation by dripping instabilities in viscous fluids. *International Journal of Multiphase Flow,* **25**, 307-320.

Bertani, R. & Cappetti, G., (1995). *Numerical simulation of the Monteverdi zone (western border of the Larderello geothermal field)*, Vol. 3. Italy, International Geothermal Association.

Bienkowski, R., Torres-Alvarado, I. S. & Hinderer, M. (2005) Geochemical modeling of acid fluids in Los Humeros geothermal field, Mexico. Paper presented at the Proceedings World Geothermal Congress 2005, Antalya, Turkey, 2005.

Bohrson, W. A. & Spera, F. J. (2001). Energy-constrained open-system magmatic processes II: application of energy-constrained assimilation - fractional crystallization (EC-AFC) model to magmatic systems. *Journal of Petrology,* **42**, 1019-1041.

Bohrson, W. A. & Spera, F. J. (2003). Energy-constrained open-system magmatic processes IV: geochemical, thermal and mass consequences of energy-constrained recharge, assimilation and fractional crystallization (EC-RAFC). *Geochemistry Geophysics Geosystems,* **4**, 8002, doi:10.1029/2002GC000316.

Bonneville, A. & Capolsini, P. (1999). THERMIC: a 2-D finite element tool to solve conductive and advective heat transfer problems in earth sciences. *Computers & Geosciences,* **25**, 1137 - 1148.

Brandeis, G. & Jaupart, C. (1986). On the interaction between convection and crystallization in cooling magma chambers. *Earth and Planetary Science Letters,* **78** 345-361.

Brandeis, G. & Marsh, B. D. (1989). The convective liquidus in a solidifying magma chamber: a fluid dynamic investigation. *Nature,* **339**, 613-616.

Campbell, I. H. & Turner, J. S. (1985). Turbulent mixing between fluids with different viscosities. *Nature,* **313**, 39-42.

Campbell, I. H. & Turner, J. S. (1986). The infuence of viscosity on fountains in magma chambers. *Journal of Petrology,* **27**, 1-30.

Campos E., J. O. & Durán M., F. (1986). Determinación preliminar del campo de temperaturas en Los Humeros, Pue. *Geotermia Revista Mexicana de Geoenergía,* **2**, 141-152.

Campos-Enríquez, J. O. & Arredondo-Fragoso, J. J. (1992). Gravity study of Los Humeros caldera complex, Mexico: structure and associated geothermal system. *Journal of Volcanology and Geothermal Research,* **49**, 69-90.

Campos-Enríquez, J. O., Domínguez-Méndez, F., Lozada-Zumaets, M., Morales-Rodríguez, H. F. & Andaverde-Arredondo, J. A. (2005). Application of the Gauss theorem to the study of silicic calderas: the calders of La Primavera, Los Azufres, and Los Humeros (Mexico). *Journal of Volcanology and Geothermal Research,* **147**, 39-67.

Cao, S., Lerche, I. & Hermanrud, C. (1988). Formation temperature estimation by inversion of borehole measurements. *Geothermics,* **53**, 979-988.

Carrasco-Núñez, G. & Branney, M. J. (2005). Progresive assembly of a massive layer of ignimbrite with a normal-to-reverse compositional zoning: the Zaragoza ignimbrite of central Mexico. *Bulletin Volcanologique,* **68**, 3-20.

Cashman, K. V. & Bergantz, G. W. (1991). Magmatic processes. *Reviews of Geophysics,* Supplement,500-512.

Castillo-Román, J., Verma, S. P. & Andaverde, J. (1991). Modelación de temperaturas bajo la caldera de Los Humeros, Puebla, México, en términos de profundidad de la cámara magmática. *Geofísica Internacional,* **30**, 149-172.

Contreras L., E., Domínguez A., B. & Rivera M., O. (1990). Mediciones petrofísicas en núcleos de perforación del campo geotérmico Los Humeros. *Geotermia Revista Mexicana de Geoenergía,* **6**, 9-42.

Currie, I. G., (1993). *Fundamental mechanics of fluids*. Boston, McGraw Hill.

Del Moro, A., Puxeddu, M., Radicati di Brozolo, F. & Villa, I. M. (1982). Rb-Sr and K-Ar ages on minerals at temperatures of 300°-400°C from deep wells in the Larderello geothermal field (Italy). *Contributions to Mineralogy and Petrology,* **81**, 340-349.

Delaney, P. T. (1988). Fortran 77 programs for conductive cooling of dikes with temperatures-dependent thermal properties and heat of crystallization. *Computers & Geosciences,* **14**, 181-212.

Deming, D. (1989). Application of bottom-hole temperature corrections in geothermal studies. *Geothermics,* **18**, 775-786.

Deming, D. & Chapman, D. S. (1988). Inversion of bottom hole temperature data: the Pineview field, Utah-Wyoming thrust belt. *Geophysics,* **53**, 707-720.

Denlinger, R. P. & Riley, F. (1984). Deformation of Long Valley caldera, Mono County, California, from 1975 to 1982. *Journal of Geophysical Research,* **89**, 8303-8314.

DiPippo, R., (2005). *Geothermal power plant: principles, applications and case studies.* Oxford, U.K., Elsevier.

Dobson, P. F. & Mahood, G. A. (1985). Volcanic stratigraphy of the Los Azufres geothermal area, Mexico. *Journal of Volcanology and Geothermal Research,* **25**, 273-287.

Drury, M. J. (1984). An a possible source of error in extracting equilibrium formation temperatures from borehole BHT data. *Geothermics,* **13**, 175-180.

Dubois, M., Flores-Marquez, L. & Royer, J. J. (1995). Numerical evaluation of pressure and temperature effects on thermal conductivity: implications for crustal geotherms. *Geofísica Internacional,* **34**, 377-384.

Economides, M. J. (1985). Material balance depletion models for two-phase and vapor-dominated geothermal reservoirs. *Geothermics,* **14**, 511-516.

Farrar, C. D., Sorey, M. L., Roeloffs, E., Galloway, D. L., Howle, J. F. & Jacobson, R. (2003). Inferences on the hydrothermal system beneath the resurgent dome in Long Valley caldera, east-central California, USA, from recent pumping tests and geochemical sampling. *Journal of Volcanology and Geothermal Research,* **127**, 305-328.

Ferriz, H. (1982) Geologic and preliminary reservoir data on the Los Humeros geothermal system, Puebla, Mexico. Paper presented at the Proceedings Eighth Workshop Geothermal Reservoir Engineering, Stanford University, Stanford, California, 1982.

Ferriz, H. (1985). Zoneamiento composicional y mineralógico en los productos eruptivos del centro volcánico de Los Humeros, Puebla, México. *Geofísica Internacional,* **24**, 97-157.

Ferriz, H. & Mahood, G. A. (1984). Eruption rates and compositional trends at Los Humeros volcanic center, Puebla, Mexico. *Journal of Geophysical Research,* **89**, 8511-8524.

Ferriz, H. & Mahood, G. A. (1987). Strong compositional zonation in a silicic magmatic system: Los Humeros, Mexican Neovolcanic Belt. *Journal of Petrology,* **28**, 171-209.

Fischer, M., Roller, K., Kuster, M., Stockhert, B. & Mcconnell, V. S. (2003). Open fissure mineralization at 2600 m depth in Long Valley exploratory well (California) - insight into the history of the hydrothermal system. *Journal of Volcanology and Geothermal Research,* **127**, 347-363.

Flores L., C., Álvarez, R., Singh, S. K. & Urrutia, J. (1978). Aeromagnetic survey of Los Humeros caldera, Mexico. *Geofísica Internacional,* **17**, 415-428.

Folch, A., Codina, R. & Marti, J. (2001). Numerical modeling of magma withdrawal during explosive caldera-forming eruptions. *Journal of Geophysical Research,* **106**, 16163 - 16175.

Folch, A. & Marti, J. (1998). The generation of overpressure in felsic magma chambers by replenishment. *Earth and Planetary Science Letters,* **163**, 301-314.

Fomin, S., Chugunov, V. & Hashida, T. (2003). Analytical modelling of the formation temperature stabilization during the borehole shut-in period. *Geophysical Journal International,* **155**, 469-478.

Fountain, J. C., Hodge, D. S. & Shaw, R. P. (1989). Melt segregation in anatectic granites: a thermo-mechanical model. *Journal of Volcanology and Geothermal Research,* **39**, 279-296.

Fournier, R. O., Sprey, M. L., Mariner, R. H. & Truesdell, A. H. (1979). Chemical and isotopic prediction of aquifer temperatures in the geothermal system at Long Valley, California. *Journal of Volcanology and Geothermal Research,* **5**, 17-34.

García-Valladares, O., Sánchez-Upton, P. & Santoyo, E. (2006). Numerical modeling of flow processes inside geothermal wells: an approach for predicting production characteristics with uncertainties. *Energy Conversion Management,* **47**, 1621-1643.

Ghiorso, M. S. (1985). Chemical mass transfer in magmatic processes. I. Thermodynamic relations and numerical algorithms. *Contributions to Mineralogy and Petrology,* **90**, 107-120.

Ghiorso, M. S. (1991). Thermodynamics of minerals and melts. *Reviews of Geophysics,* Supplement,446-456.

Ghiorso, M. S. & Carmichael, I. S. E. (1985). Chemical mass transfer in magmatic processes. II. Aplications in equilibrium crystallization, fractionation and assimilation. *Contributions to Mineralogy and Petrology,* **90**, 121-141.

Ghiorso, M. S. & Sack, R. O. (1995). Chemical mass transfer in magmatic processes IV. A revised and internally consistent thermodynamic model for the interpolation and extrapolation of liquid-solid equilibria in magmatic systems at elevated temperatures and pressures. *Contributions to Mineralogy and Petrology,* **119**, 197-212.

Giberti, G. & Sartoris, G. (1989). Evaluation of approximations in modeling the thermal history of a volcanic area. *Journal of Volcanology and Geothermal Research,* **36**, 233-240.

Giberti, G., Moreno, S. & Sartoris, G. (1984a). Evaluation of approximations in modelling the cooling of magmatic bodies. *Journal of Volcanology and Geothermal Research,* **20**, 297-310.

Giberti, G., Moreno, S. & Sartoris, G. (1984b). Thermal history of Phlegraean fields (Italy) in the last 50,000 years: a schematic numerical model. *Bulletin Volcanologique,* **47**, 331-341.

González-Morán, T. & Suro-Pérez, V. (1982). Obtención de parámetros físicos de las estructuras principales localizadas en la caldera de los Humeros, Puebla. *Geofísica Internacional,* **21**, 57-68.

González Partida, E., Carrillo-Chávez, A., Levresse, G., Tello-Hinojosa, E., Venegas-Salgado, S., Ramírez-ilva, G., Pal-Verma, M., Tritlla, J. & Camprubí, A. (2005). Hydro-geochemical and isotopic fluid evolution of the Los Azufres geothermal field, Central Mexico. *Applied Geochemistry,* **20**, 23-39.

Grunder, A. L. & Mahood, G. A. (1988). Physical and chemical models of zoned silicic magmas: the Loma Seca tuff and Calabozos caldera, southern Andes. *Journal of Petrology,* **29**, 831-867.

Guglielminetti, M. (1985). Electric power production from geothrmal energy. *Geothermics,* **14**, 157-163.

Hansen, U. & Yuen, D. A. (1990). Nonlinear physics of double-diffusive convection in geological systems. *Earth-Science Reviews,* **29**, 385-399.

Hasan, A. R. & Kabir, C. S. (1991). Heat transfer during two-phase flow in wellbores: Part II - Wellbore fluid temperature. *Society of Petroleum Engineers, SPE22948*, 695-708.

Hasan, A. R. & Kabir, C. S. (1994). Static reservoir temperature determination from transient data after mud circulation. *Society of Petroleum Engineers Drilling & Completion,* March, 17-24.

Hermance, J. F., Slocum, W. M. & Neumann, G. A. (1984). The Long Valley/Mono Basin volcanic complex: a preliminary magnetotelluric and magnetic variation interpretation. *Journal of Geophysical Research,* **89**, 8325-8337.

Hermanrud, C., Cao, S. & Lerche, I. (1990). Estimates of virgin rock temperature derived from BHT measurements: bias and error. *Geophysics,* **55**, 924-931.

Hill, D. P. (1984). Monitoring unrest in a large silicic caldera, the Long Valley-Inyo craters volcanic complex in east-central California. *Bulletin Volcanologique,* **47**, 371-395.

Hill, D. P., Langbein, J. O. & Prejean, S. (2003). Relations between seismicity and deformation during unrest in Long Valley caldera, California, from 1995 through 1999. *Journal of Volcanology and Geothermal Research,* **127**, 175-193.

Hiriart, G. & Andaluz, J. I. (2000) Strategies and economics of geothermal power development in Mexico. Paper presented at the Proceedings of the World Geothermal Congress 2000, Kyushu-Tohoku, Japan, 2000.

Hirschmann, M. M., Ghiorso, M. S., Wasylenki, L. E., Asimow, P. D. & Stolper, E. M. (1998). Calculation of peridotite partial melting from thermodynamic models of minerals and melts. I. Review of methods and comparison with experiments. *Journal of Petrology,* **39**, 1091-1115.

Hort, M. (1997). Cooling and crystallization in sheet-like magma bodies revisited. *Journal of Volcanology and Geothermal Research,* **76**, 297-317.

Howle, J. F., Langbein, J. O., Farrar, C. D. & Wilkinson, S. K. (2003). Deformation near the Casa Diablo geothermal well field and related processes Long Valley caldera, Eastern California, 1993-2000. *Journal of Volcanology and Geothermal Research,* **127**, 365-390.

Huppert, H. E. & Sparks, R. S. J. (1984). Double-diffusive convection due to crystallization in magmas. *Annual Review of Earth and Planetary Sciences,* **12**, 11-37.

Huppert, H. E. & Sparks, R. S. J. (1988). Melting the roof of a chamber containing a hot, turbulently convecting fluid. *Journal of Fluid Mechanics,* **188**, 107-131.

Huppert, H. E. & Woods, A. W. (1988). The role of volatiles in magma chamber dynamics. *Nature,* **420**, 483-485?

Huppert, H. E., Sparks, R. S. J. & Turner, J. S. (1982). Effects of volatiles on mixing in calc-alkaline magma systems. *Nature,* **297**, 554 - 557.

Huppert, H. E., Stephen, R., Sparks, J., Whitehead, J. A. & Hallworth, M. A. (1986). Replenishment of magma chambers by light inputs. *Journal of Geophysical Research,* **91**, 6113-6122.

Ida, Y. (1995). Magma chamber and eruptive processes at Izu-Oshima volcano, Japan: buoyancy control of magma migration. *Journal of Volcanology and Geothermal Research,* **66**, 53-67.

Jaupart, C. & Tait, S. (1995). Dynamics of differentiation in magma reservoirs. *Journal of Geophysical Research,* **100**, 17615-17636.

Jellinek, A. M. & DePaolo, D. J. (2003). A model for the origin of large silicic magma chambers: precursors of caldera-forming eruptions. *Bulletin of Volcanology,* **65**, 363-381.

Johnson, M. C. & Rutherford, M. J. (1989). Experimental calibration of the aluminum-in-hornblende geobarometer with application to Long Valley caldera (California) volcanic rocks. *Geology,* **17**, 837-841.

Kaneko, K. & Koyaguch, T. (2004). Experimental study on the effects of crustal temperature and composition on assimilation with fractional crystallization at the floor of magma chambers. *Journal of Volcanology and Geothermal Research,* **129**, 155-172.

Kojitani, H. & Akaogi, M. (1997). Melting enthalpies of mantle peridotite: calorimetric determinations in the system CaO-MgO-Al2O3-SiO2 and application to magma generation. *Earth and Planetary Science Letters,* **154**, 209-222.

Koyaguchi, T., Hallworth, M. A., Huppert, H. E. & Sparks, R. S. J. (1990). Sedimentation of particles from a convecting fluid. *Nature,* **343**, 447-450.

Kuritani, T. (2004). Magmatic differentiation examined with a numerical model considering multicomponent thermodynamics and momentum, energy and species transport. *Lithos,* **74**, 117-130.

Kutasov, I. M. (2003). Dimensionless temperature at the wall of an infinite long cylindrical source with a constant heat flow rate. *Geothermics,* **32**, 63 - 68.

Kutasov, I. M. & Eppelbaum, L. V. (2005). Determination of formation temperature from bottom-hole temperature logs - a generalized Horner method. *Journal of Geophysical Engineering,* **2**, 90-96.

Kutasov, I. M., Caruthers, R. M., Targhi, A. K. & Chaaban, H. M. (1988). Prediction of downhole circulating and shut-in temperatures. *Geothermics,* **17**, 607-618.

Lachenbruch, A. H., Sass, J. H., Munroe, R. J. & Moses, T. H. (1976). Geothermal setting and simple heat conduction models for the Valles caldera. *Journal of Geophysical Research,* **81**, 769-784.

Lalou, C., Reyss, J. L. & Brichet, E. (1993). Actinide-series disequilibrium as a tool to establish the chronology of deep-sea hydrothermal activity. *Geochimica et Cosmochimica Acta,* **57**, 1221-1231.

Langbein, J. O. (2003). Deformation of the Long Valley caldera, California: inferences from measurements from 1988 to 2001. *Journal of Volcanology and Geothermal Research,* **127**, 247-267.

López Mendiola, J. M. & Munguía B., F. (1989). Evidencias geoquímicas del fenómeno de ebullición en el campo de Los Humeros, Puebla, México. *Geotermia Revista Mexicana de Geoenergía,* **5**, 89-106.

López Mendiola, J. M. & Munguía Bracamontes, F. (1987). Configuración de isovalores de temperatura para el campo geotérmico de Los Humeros, Puebla, México. *Geotermia Revista Mexicana de Geoenergía,* **3**, 255-261.

Lowell, R. P. (1985). Double-diffusive convection in partially molten silicate systems: its role during magma production and in magma chambers. *Journal of Volcanology and Geothermal Research,* **26**, 1-24.

Lowell, R. P. & Burnell, D. K. (1991). Mathematical modeling of conductive heat transfer from freezing, convecting magma chamber to a single-pass hydrothermal system: implications for seafloor black smokers. *Earth and Planetary Science Letters,* **104**, 59 - 69.

Luetgert, J. H. & Mooney, W. D. (1985). Crustal refraction profile of the long valley caldera, California, from the january 1983 mammoth lakes earthquake swarm. *Bolletin of the Seismological Society of America,* **75**, 211 - 221.

Luheshi, M. N. (1983). Estimation of formation temperature from borehole measurements. *Geophysical Journal of the Royal Astronomical Society,* **74**, 747-776.

Luhr, J. F. & Williams, S. N. (1991). Volcanology. *Reviews of Geophysics,* Supplement, 513-531.

Mahood, G. A. (1981a). Chemical evolution of a Pleistocene rhyolitic center: Sierra La Primavera, Jalisco, México. *Contributions to Mineralogy and Petrology,* **77**, 129-149.

Mahood, G. A. (1981b). A summary of the geology and petrology of the Sierra La Primavera, Jalisco, Mexico. *Journal of Geophysical Research,* **86**, 10137-10152.

Mahood, G. A. (1984). Pyroclastic rocks and calderas associated with strongly peralkaline magmatism. *Journal of Geophysical Research,* **89**, 8540-8552.

Mahood, G. A. & Halliday, A. N. (1988). Generation of high-silica rhyolite: a Nd, Sr, and O isotopic study of Sierra La Primavera, Mexican Neovolcanic Belt. *Contributions to Mineralogy and Petrology,* **100**, 183-191.

Mahood, G. A., Truesdell, A. H. & Templos M., L. A. (1983). A reconnaissance geochemical study of La Primavera geothermal area, Jalisco, Mexico. *Journal of Volcanology and Geothermal Research,* **16**, 247-261.

Márquez, A., Verma, S. P., Anguita, F., Oyarzun, R. & Brandle, J. L. (1999). Tectonics and volcanism of Sierra Chichinautzin: extension at the front of the Central Trans-Mexican Volcanic Belt. *Journal of Volcanology and Geothermal Research,* **93**, 125-150.

Márquez, A., Oyarzun, R., de Ignacio, C. & Doblas, M. (2001). Southward migration of volcanic activity in the central Mexican Volcanic Belt: asymmetric extension within a two-layer crustal stretching model. *Journal of Volcanology and Geothermal Research,* **112**, 175-187.

Marsh, B. D. (1989). Magma chambers. *Annual Reviews of Earth and Planetary Sciences,* **12**, 439-474.

Marsh, B. D. (1996). The 1995 Hallimond Lecture. Solidification fronts and magmatic evolution. *Mineralogical Magazine,* **60**, 5-40.

Marsh, J. (1989). Geochemical constraints on coupled assimilation and fractional crystallization involving upper crustal compositions and continental thoeliitic magma. *Earth and Planetary Science Letters,* **92**, 70-80.

Martin, D. M., Griffiths, R. W. & Campbell, I. H. (1987). Compositional and thermal convection in magma chambers. *Contributions to Mineralogy and Petrology,* **96**,465-475.

Martínez Serrano, R. G. & Alibert, C. (1994). Características geoquímicas de las rocas volcánicas del sistema geotérmico Los Humeros, Puebla y su relación con la mineralogía de alteración. *Geofísica Internacional,* **33**, 585-605.

Mastin, L. G. (2002). Insights into volcanic conduit flow from an open-source numerical model. *Geochemistry, Geophysics, Geosystems,* **3**, 1-18.

McConnell, V. S., Shearer, C. K., Eichelberger, J. C., Keskinen, M. J., Layer, P. W. & Papike, J. J. (1995). Rhyolite intrusions in the intracaldera Bishop Tuff, Long Valley Caldera, California. *Journal of Volcanology and Geothermal Research,* **67**, 41-60.

McConnell, V. S., Valley, J. W. & Eichelberger, J. C. (1997). Oxygen isotope compositions of intracaldera rocks: hydrothermal history of the Long Valley Caldera, California. *Journal of Volcanology and Geothermal Research,* **76**, 83-109.

McLeod, P. (1999). The role of magma buoyancy in caldera-forming eruptions. *Geophysical Research Letters,* **26**, 2299-2302.

Medina Martínez, M. (2000). Intervalos productores de fluidos en los pozos H'35 y H'39 del campo geotérmico Los Humeros, Puebla, México. *Geotermia Revista Mexicana de Geoenergía,* **16**, 51-60.

Meidav, T., (1995). *Approximation of geothermal reservoir characteristics based on gravimetry*, Vol. 2. Italy, International Geothermal Association.

Mena, M. & González-Morán, T. (1978). Regional gravity of Los Humeros volcanic area. *Geofísica Internacional,* **17**, 429-443.

Mourtada-Bonnefoi, C. C., Provost, A. & Albarède, F. (1999). Thermochemical dynamics of magma chambers: a simple model. *Journal of Geophysical Research,* **104**, 7103-7115.

Moya, S. L., Ruiz, J. N., Aragón, A. & Iglesias, E. (1995). Modelo numérico en diferencias finitas para el transporte de masa y energía en yacimientos geotérmicos con bióxido de carbono. *Geotermia Revista Mexicana de Geoenergía,* **11**, 37-51.

Moya, S. L., Aragón, A., Iglesias, E. R. & Santoyo, E. (1998). Prediction of mass deliverability from a single wellhead measurement and geothermal inflow performance reference curves. *Geothermics,* **27**, 317-329.

Ni, J. & Beckermann, C. (1991). A volume-averaged two-phase model for transport phenomena during solidification. *Metallurgical Transactions,* **22**B, 349-361.

Ni, J. & Incropera, F. P. (1995a). Extension of the continuum model for transport phenomena occurring during metal alloy sodification - I. The conservation equations. *International Journal of Heat and Mass Transfer,* **38**, 1271-1284.

Ni, J. & Incropera, F. P. (1995b). Extension of the continuum model for transport phenomena occurring during metal alloy sodification - II. Microscopic considerations. *International Journal of Heat and Mass Transfer,* **38**, 1285-1296.

Nicholls, J. & Stout, M. Z. (1982). Heat effects of assimilation, crystallization, and vesiculation in magmas. *Contributions to Mineralogy and Petrology,* **81**, 328-339.

Nielsen, R. L. (1985). A method for the elimination of the compositional dependence of trace element distribution coefficients. *Geochimica et Cosmochimica Acta,* **49**, 1775-1779.

Nielsen, R. L. (1988a). A model for the simulation of combined major and trace element liquid line of descent. *Geochimica et Cosmochimica Acta,* **52**, 27-38.

Nielsen, R. L. (1988b). Trace.for: a program for the calculation of combined major and trace-element liquid lines of descent for natural magmatic systems. *Computers & Geosciences,* **14**, 15-35.

Nielsen, R. L. (1992). BIGD.FOR: a fortran program to calculate trace-element partition coefficients for natural mafic and intermediate composition magmas. *Computers & Geosciences,* **18**, 773-788.

Nielsen, S. B., Balling, N. & Christian, H. S. (1990). Formation temperatures determined fron stochastic inversion of borehole observations. *Geophysica Journal International,* **101**, 581-590.

Oldenburg, C. M., Spera, F. J., Yuen, D. A. & Sewell, G. (1989). Dynamic mixing in magma bodies: theory, simulations, and implications. *Journal of Geophysical Research,* **94**, 9215-9236.

Oldenburg, C. M., Spera, F. J. & Yuen, D. A. (1990). Self-organization in convective magma mixing. *Earth-Science Reviews,* **29**, 331-348.

Pandarinath, K., Pushparani, D. E., Torres-Alvarado, I. & Verma, S. P. (2006). X-ray diffraction analysis of hydrothermal minerals from the Los Azufres geothermal system, Mexico. *International Geology Review,* **48**, 174-190.

Patankar, S. V., (1980). *Numerical heat transfer and fluid flow.* New York, McGraw Hill.

Pérez Reynoso, J. (1979). Geología y petrografía de la caldera de Los Humeros. *Geomimet,* **91**, 97-106.

Ponce, L. & Rodríguez, C. (1978). Microearthquake activity associated to Los Humeros caldera, Mexico: preliminary survey. *Geofísica Internacional,* **17**, 461-478.

Prensky, S. (1992). Temperature measurements in boreholes: an overview of engineering and scientific applications. *The Log Analyst,* **33**, 313-333.

Pribnow, D. F. C., Schutze, C., Hurter, S. J., Flechsig, C. & Sass, J. H. (2003). Fluid flow in the resurgent dome of Long Valley caldera: implications from thermal data and deep electrical sounding. *Journal of Volcanology and Geothermal Research,* **127**, 329-345.

Prol, R. M. & González-Morán, T. (1982). Modelo preliminar del régimen térmico conductivo en la caldera de Los Humeros, Puebla. *Geofísica Internacional,* **21**, 295-307.

Quick, J. E., Sinigoi, S. & Mayer, A. (1994). Emplacement dynamics of a large mafic intrusion in the lower crust, Ivrea-Verbano Zone, northern Italy. *Journal of Geophysical Research,* **99**, 21559-21573.

Richet, P. & Bottinga, Y. (1984). Anorthite, andesine, wollastonite, diopside, cordierite and pyrope: thermodynamics of melting, glass transitions, and properties of the amorphous phases. *Earth and Planetary Science Letters,* **67**, 415-432.

Rihs, S., Condomines, M. & Sigmarsson, O. (2000). U, Ra and Ba incorporation during precipitation of hydrothermal carbonates: implications for ^{226}Rn-Ba dating of impure travertines. *Geochimica et Cosmochimica Acta,* **64**, 661-671.

Roeloffs, E., Sneed, M., Galloway, D. L., Sorey, M. L., Farrar, C. D., Howle, J. F. & Hughes, J. (2003). Water-level changes induced by local and distant earthquakes at Long Valley caldera, California. *Journal of Volcanology and Geothermal Research,* **127**, 269-303.

Royer, J. & Flores, L. (1994). Two-dimensional natural convection in an anisotropic and heterogeneus porous medium with internal heat generation. *International Journal of Heat and Mass Transfer,* **37**, 1387-1399.

Rudman, M. (1992). Two-phase natural convection: implications for crystal settling in magma chambers. *Physics of the Earth and Planetary Interiors,* **72**, 153-172.

Sampson, D. E. (1987). Textural heterogeneities and vent area structures in the 600-year-old lavas of the Inyo volcanic chain, eastern California. *Geological Society of America Bulletin,* **212**, 89-101.

Sampson, D. E. & Cameron, K. L. (1987). The geochemistry of the Inyo volcanic chain: multiple magma systems in the Long Valley region, eastern California. *Journal of Geophysical Research,* **92**, 10403-10421.

Sánchez Upton, P. (1994). Producción de vapor en pozos geotérmicos. *Geotermia Revista Mexicana de Geoenergía,* **10**, 39-44.

Sanders, C. O. (1984). Location and configuration of magma bodies beneath Long Valley, California determined from anomalous earthquake signals. *Journal of Geophysical Research,* **89**, 8287-8302.

Santoyo, E., Garcia, A., Espinosa, G., Hernandez, I. & Santoyo, S. (2000). STATIC_TEMP: a useful computer code for calculating static formation temperatures in geothermal wells. *Computers & Geosciences,* **26**, 201-217.

Savage, J. C. & Cockerham, R. S. (1984). Earthquake swarm in Long Valley caldera, California, January 1983: evidence for dike inflation. *Journal of Geophysical Research,* **89**, 8315-8324.

Shen, P. Y. & Beck, A. E. (1986). Stabilization of bottom hole temperature with finite circulation time and fluid flow. *Geophysical Journal of the Royal Astronomical Society,* **86**, 63-90.

Simakin, A. & Botcharnikov, R. (2001). Degassing of stratified magma by compositional convection. *Journal of Volcanology and Geothermal Research,* **105**, 207-224.

Snyder, D. & Tait, S. (1995). Replenishment of magma chambers: comparison of fluid-mechanic experiments with field relations. *Contributions to Mineralogy and Petrology,* **122**, 230-240.

Sorey, M. L., Mcconnell, V. S. & Roeloffs, E., (2003). *Special Issue on Crustal unrest in Long Valley caldera, California: new interpretation from geophysical and hydrologic monitoring and deep drilling,* **127**, 165-392.

Spera, F. J. (1980). Thermal evolution of plutons: a parametrized approach. *Science,* **207**, 299 - 301.

Spera, F. J. (2000) Physical properties of magma. In *Encyclopedia of Volcanoes*, pp. 171-190. Academic Press.

Spera, F. J. & Bohrson, W. A. (2001). Energy-constrained open-system magmatic processes I: general model and energy-constrained assimilation and fractional crystallization (EC-AFC) formulation. *Journal of Petrology,* **42**, 999-1018.

Spera, F. J. & Bohrson, W. A. (2002). Energy-constrained open-system magmatic processes 3: energy-constrained recharge, assimilation and fractional crystallization (EC-RAFC). *Geochemistry Geophysics Geosystems,* **3**, 8001, doi:10.1029/2002GC000315.

Spera, F. J. & Crisp, J. A. (1981). Eruption volume, periodicity, and caldera area: relationships and inferences on development of compositional zonation in silicic magma chamber. *Journal of Volcanology and Geothermal Research,* **11**, 169-187.

Spera, F. J., Oldenburg, C. M., Christensen, C. & Todesco, M. (1995). Simulations of convection with crystallization in the system KAlSi2O6-CaMgSi2O6: implications for compositionally zoned magma bodies. *American Mineralogist,* **80**, 1188-1207.

Spera, F. J., Yuen, D. A. & Kirschvink, S. J. (1982). Thermal boundary layer convection in silicic magma chambers: effects of temperature-dependent rheology and implications for thermogravitational chemical fractionation. *Journal of Geophysical Research,* **87**, 8755-8767.

Stimac, J. A., Goff, F. & Wohletz, K. (2001). Thermal modeling of the Clear Lake magmatic-hydrothermal system, California, USA. *Geothermics,* **30**, 349-390.

Stix, J. & Gorton, M. P. (1993). Replenishment and crystallization in epicontinental silicic magma chambers: evidence from the Bandelier magmatic system. *Journal of Volcanology and Geothermal Research,* **55**, 201-215.

Sturchio, N. C. & Binz, C. M. (1988). Uranium-series age determination of calcite veins, VC-1 drill core, Valles Caldera, New Mexico. *Journal of Geophysical Research,* **93**, 6097-6102.

Sturchio, N. C., Murrell, M. T., Pierce, K. L. & Sorey, M. L. (1992) Yellowstone travertines: U-series ages and isotope ratios (C, O, Sr, U). Paper presented at the 7th International Symposium on water-rock interaction, Park City, Utah, USA, 1992.

Tait, S. R. (1988). Samples from the crystallising boundary layer of a zoned magma chamber. *Contributions to Mineralogy and Petrology,* **100**, 470-483.

Tan, E. (1985). Reservoir characteristics of the Kizildere geothermal field. *Geothermics,* **14**, 419-428.

Tello Hinojosa, E. (1992a). Características geoquímicas e isotópicas de los fluidos producidos por los pozos de Los Humeros, Puebla, México. *Geotermia Revista Mexicana de Geoenergía,* **8**, 3-48.

Tello Hinojosa, E. (1992b). Composición química de la fase líquida a descarga total y a condiciones de reservorio de pozos geotérmicos de Los Humeros, Puebla, México. *Geofísica Internacional,* **31**, 383-390.

Tiampo, K. F., Rundle, J. B., Fernandez, J. & Langbein, J. O. (2000). Spherical and ellipsoidal volcanic sources at Long Valley caldera, California, using a genetic algorithm inversion technique. *Journal of Volcanology and Geothermal Research,* **102**, 189-206.

Torres R., M. A. (1988). Aplicación de la función de derivada en el análisis de pruebas de presión en los pozos del campo geotérmico de Los Humeros, Puebla, México. *Geotermia Revista Mexicana de Geoenergía,* **4**, 203-211.

Torres R., M. A., (1995). *Characterization of the reservoir of the Los Humeros, Mexico geothermal field*, Vol. 3. Italy, International Geothermal Association.

Torres-Alvarado, I. S. & Satir, M. (1998). Geochemistry of hydrothermally altered rocks from Los Azufres geothermal field, Mexico. *Geofísica Internacional,* **37**, 201-213.

Trial, A. F. & Spera, F. J. (1990). Mechanisms for the generation of compositional heterogeneities in magma chambers. *Geological Society of America Bulletin,* **102**, 353-367.

Troise, C., Castagnolo, D., Peluso, F., Gaeta, F., G., M. & Natale, G. (2001). A 2D mechanical-thermalfluid-dynamical model for geothermal systems at calderas: an aplication to campi flegrei, italy. *Journal of volcanologyu and geothermal rsearch,* **109**, 1-12.

Turner, J. S., Huppert, H. E. & Sparks, R. S. (1983). An experimental investigation of volatile exsolution in evolving magma chamber. *Journal of Volcanology and Geothermal Research,* **16**, 263 - 277.

Valentine, G. A., (1992). *Magma chamber dynamics*. In: Encyclopedia of Earth System *Science*, **3**, 1-17, Academic Press.

Valentine, G. A. (1994). Multifield governing equations for magma dynamics. *Geophysical and Astrophysical Fluid Dynamics,* **78**, 193-210.

Valentine, G. A., Zhang, D. & Robinson, B. A. (2002). Modeling complex, nonlinear geological processes. *Annual Reviews of Earth and Planetary Sciences,* **30**, 35-64.

van der Bogaard, P. & Schirnick, C. (1995). $^{40}Ar/^{39}Ar$ laser probe ages of Bishop Tuff quartz phenocrysts substantiate long-lived silicic magma chamber at Long Valley, United States. *Geology,* **23**, 759-762.

Velasco-Tapia, F. & Verma, S. P. (2001). First partial melting inversion model for a rift-related origin of the Sierra de Chichinautzin volcanic field, central Mexican Volcanic Belt. *International Geology Review,* **43**, 788-817.

Verma, M. P., Verma, S. P. & Sanvicente, H. (1990). Temperature field simulation with stratification model of magma chamber under Los Humeros caldera, Puebla, Mexico. *Geothermics,* **19**, 187-197.

Verma, S. P. (1983). Magma genesis and chamber processes at Los Humeros caldera, Mexico - Nd and Sr isotope data. *Nature,* **301**, 52-55.

Verma, S. P. (1984). Alkali and alkaline earth element geochemistry of Los Humeros caldera, Puebla, Mexico. *Journal of Volcanology and Geothermal Research,* **20**, 21-40.

Verma, S. P. (1985). On the magma chamber characteristics as inferred from surface geology and geochemistry: examples from Mexican geothermal areas. *Physics of the Earth and Planetary Interiors,* **41**, 207-214.

Verma, S. P. (1994). Geochemical and isotopic constraints on the origin of mafic volcanism in central Mexico. *Mineralogical Magazine,* **58**A, 938-939.

Verma, S. P. (1999a). Preface special issue on "Rift-related volcanism: geology, geochemistry, and geophysics". *Journal of Volcanology and Geothermal Research,* **93**, vii-ix.

Verma, S. P. (1999b). Geochemistry of evolved magmas and their relationship to subduction-unrelated mafic volcanism at the volcanic front of the central Mexican Volcanic Belt. *Journal of Volcanology and Geothermal Research,* **93**, 151-171.

Verma, S. P. (2000). Geochemical evidence for a lithospheric source for magmas from Los Humeros caldera, Puebla, Mexico. *Chemical Geology,* **164**, 35-60.

Verma, S. P. (2001). Geochemical evidence for a rift-related origin of bimodal volcanism at Meseta Río San Juan, North-Central Mexican Volcanic Belt. *International Geology Review,* **43**, 475-493.

Verma, S. P. (2002a) Optimisation of the exploration and evaluation of geothermal resources. In *Geothermal Energy Resources for Developing Countries*, Chandrasekharam, D. and Bundschuh, J. eds, pp. 195-224. Swets & Zeitlinger, B.V., A.A. Balkena Publishers, Rotherdam.

Verma, S. P. (2002b). Absence of Cocos plate subduction-related basic volcanism in southern Mexico: a unique case on Earth? *Geology,* **30**, 1095-1098.

Verma, S. P. (2003). Geochemical and Sr-Nd isotopic evidence for a rift-related origin of magmas in Tizayuca volcanic field, Central Mexican Volcanic Belt. *Journal of the Geological Society of India,* **61**, 257-276.

Verma, S. P. (2004). Solely extension-related origin of the eastern to west-central Mexican Volcanic Belt (Mexico) from partial melting inversion model. *Current Science,* **86**, 713-719.

Verma, S. P., (2005). *Estadística básica para el manejo de datos experimentales: aplicación en la geoquímica (Geoquimiometría)*. México, D.F., UNAM.

Verma, S. P. (2006). Extension related origin of magmas from a garnet-bearing source in the Los Tuxtlas volcanic field, Mexico. *International Journal of Earth Sciences,* **95**, 871-901.

Verma, S. P. & Andaverde, J. (1995) *Temperature field distribution from cooling of a magma chamber*, Florence, 1995.

Verma, S. P. & Andaverde, J. (1996). Temperature distributions from cooling of a magma chamber in Los Azufres geothermal field, Michoacán, Mexico. *Geofísica Internacional,* **35**, 105-113.

Verma, S. P. & Hasenaka, T. (2004). Sr, Nd, and Pb isotopic and trace element geochemical contraints for a veined-mantle source of magmas in the Michoacán-Guanajuato volcanic field, west-central Mexican Volcanic Belt. *Geochemical Journal,* **38**, 43-65.

Verma, S. P. & Lopez M., M. (1982). Geochemistry of Los Humeros caldera, Puebla, Mexico. *Bulletin Volcanologique,* **45**, 63-79.

Verma, S. P. & Rodríguez-González, U. (1997). Temperature field distribution from cooling of a magma chamber in La Primavera Caldera, Jalisco, Mexico. *Geothermics,* **26** 25-42.

Verma, S. P., Torres-Alvarado, I. S. & Sotelo-Rodríguez, Z. T. (2002). SINCLAS: standard igneous norm and volcanic rock classification system. *Computers & Geosciences,* **28**, 711-715.

Verma, S. P., Torres-Alvarado, I. S. & Velasco-Tapia, F. (2003). A revised CIPW norm. *Schweizerische Mineralogische und Petrographische Mitteilungen,* **83**, 197-216.

Verma, S. P., Torres-Alvarado, I. S., Satir, M. & Dobson, P. F. (2005). Hydrothermal alteration effects in geochemistry and Sr, Nd, Pb, and O isotopes of magmas from the Los Azufres geothermal field (Mexico): a statistical approach. *Geochemical Journal,* **39**, 141-163.

Verma, S. P., Pandarinath, K., Santoyo, E., González-Partida, E., Torres-Alvarado, I. & Tello-Hinojosa, E. (2006a). Fluid chemistry and temperature prior to exploitation at the Las Tres Vírgenes geothermal field, Mexico. *Geothermics,* **35**, 156-180.

Verma, S. P., Andaverde, J. & Santoyo, E. (2006b). Statistical evaluation of methods for the calculation of static formation temperatures in geothermal and oil wells using an extension of the error propagation theory. *Journal of Geochemical Exploration,* **89**, 398-404.

Verma, S. P., Andaverde, J. & Santoyo, S. (2006c). Application of the Error Propagation Theory in estimates of static formation temperatures in geothermal and petroleum boreholes. *Energy Conversion and Management,* **47**, 3659-3671.

Verma, S. P., Guevara, M. & Agrawal, S. (2006d). Discriminating four tectonic settings: Five new geochemical diagrams for basic and ultrabasic volcanic rocks based on log-ratio transformation of major-element data. *Journal of Earth System Science,* **115**, 485-528.

Versteeg, H. K. & Malalasekera, W., (1995). *An introduction to computational fluid dynamics The finite volume method.* Harlow, England, Pearson, Prentice Hall.

Vogel, T. A., Woodburne, T. B., Eichelberger, J. C. & Layer, P. W. (1994). Chemical evolution and periodic eruption of mafic lava flows in the west moat of Long Valley Caldera, California. *Journal of Geophysical Research,* **99**, 19829-19842.

Wallace, G. S. & Bergantz, G. W. (2002). Wavelet-based correlation (WBC) of zoned crystal populations and magma mixing. *Earth and Planetary Science Letters,* **202**, 133-145.

Weinberg, R. F. & Leitch, A. M. (1998). Mingling in mafic magma chambers replenished by light felsic inputs: fluid dynamical experiments. *Earth and Planetary Science Letters,* **157**, 41-56.

Wilhelm, H. (1990). A new approach to the borehole temperature relaxation method. *Geophysical Journal International,* **103**, 469-481.

Wilhelm, H. (2000). Undisturbed temperature in the main drillhole of the German continental deep drilling program predicted from temperature logs recorded after shut-in. *Geothermics,* **29**, 393-406.

Wohletz, K., Civetta, L. & Orsi, G. (1999). Thermal evolution of the Phlegraean magmatic system. *Journal of Volcanology and Geothermal Research,* **91**, 381-414.

Wollenberg, H. A., Flexser, S. & Smith, A. R. (1995). Mobility and depositional controls of radioelements in hydrothermal systems at the Long Valley and Valles calderas. *Journal of Volcanology and Geothermal Research,* **67**, 171-186.

In: Geothermal Energy Research Trends
Editor: Herman I. Ueckermann, pp. 163-201
ISBN: 1-60021-683-8

Chapter 7

DETERMINATION OF THE DAMAGE EFFECT IN GEOTHERMAL WELLS USING INFLOW TYPE CURVES

***A. A. Aragón*[1], *S. L. Moya*[2] *and A. M. C. Suárez*[3]**
[1]Instituto de Investigaciones Eléctricas, México
[2]CENIDET-SEP, México
[3]Univ. Michoacana de San Hidalgo, México

Abstract

The inflow curves are equivalent to the characteristic curves of production at bottom-hole conditions. They have been used for several decades as a tool for the analysis of the productivities of wells, to establish criteria for their operation. The development and application of the inflow curves begins in petroleum engineering and later on with adaptations, according to the type of fluid, they are applied to geothermal engineering.

In this chapter the equations of inflow behavior along with damage effect for oil systems are shown. The existent relationships of the inflow behavior in geothermal reservoir engineering are revised. The relationships of the inflow behavior of geothermal fluid are exposed and discussed, considering the fluid as a ternary mixture (H_2O-CO_2-$NaCl$) for low and high salinity. Those factors influencing the presence of damage are described and the parameters applicable to geothermal systems are thus obtained. Representative variables of the damage effect and the average physical characteristics of a geothermal system, within the relationships of the inflow behavior are incorporated. The acquisition of the first geothermal inflow type-curve for different values of the damage is shown.

The geothermal inflow type-curves obtained for different damage values are applicable to data from production tests in geothermal wells. They are a useful tool in determining the value of the damage in the well. The methodology implies the selection of the corresponding geothermal type-curve according to the characteristics of fluid salinity.

The methodology proposed to determine the value of the damage effect in a well is described; starting from the geothermal inflow type-curves using data from a production test. We incorporate the method of Jones *et al.* (1976) as a verification tool for the presence or absence of damage.

We show the use of this methodology with data from production tests in wells from three different fields, using the two geothermal inflow type-curves (for low and high salinity). The ascertained damage values are verified in the results of Jones' method, finding consistency in both techniques.

This methodology is useful as a tool for the diagnosis of the damage effect starting from production tests of a geothermal well. Identifying the numerical value of the damage permits the establishment of criteria for preventive and remedial operations in such wells.

Introduction

The relationships of the inflow type-curves are characteristic curves of production at bottom-hole conditions. They are built starting from the values of both pressure and flow measured during the production tests of a well, or calculating them directly from their characteristic curves of production using a well numerical simulator.

The curves of inflow (and the characteristic curves) are specific to each well and vary according to the stage of their productive life. They are also a reflection of the thermo-physical characteristics of the formation and of the properties of the fluid in the reservoir.

The schematics in Fig. 1 illustrate the behavior presented by the inflow curves of a well at different stages of exploitation. The decline exhibited in these curves is translated as a peculiar decline of the well and of the reservoir in general.

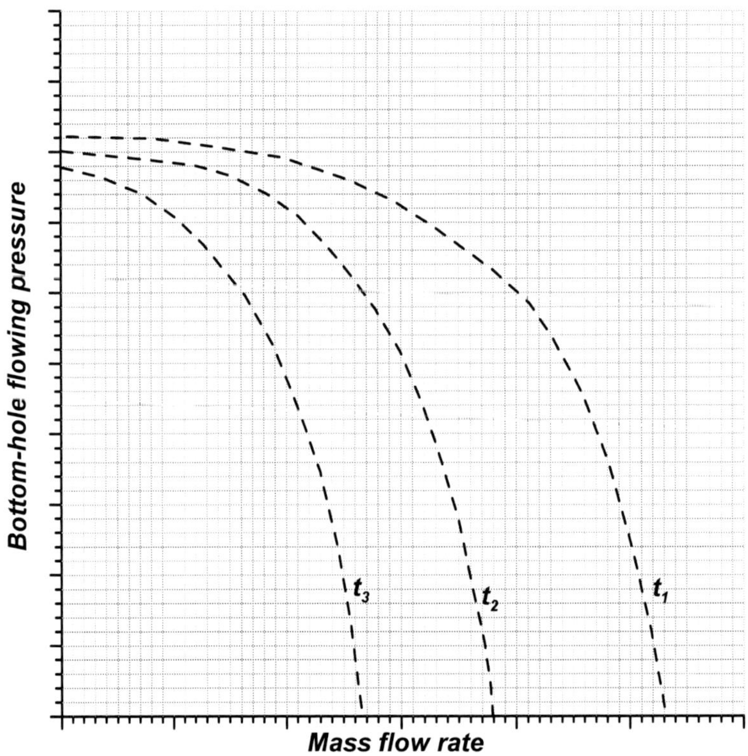

Figure 1. Typical inflow curves of a well obtained from discharge tests at different times of its productive life.

Decline within a well begins at the onset of exploitation and is a direct function of the reservoir characteristics in its surrounding formation. The tendency toward decline ordinarily shown by these wells, can adopt harmonic, exponential or hyperbolic models. The behavior of decline that a well adopts begins to be evident after a period of time in which it has been subjected to continuous exploitation. Moreover, reaction times for each well are related to the surmounting of transitory effects.

Figure 2 shows a schematic representation of the behavior and trend of the decline. The transitory effects occur at the beginning of the exploitation. For a constant diameter the well production behavior is a linear function of time. In this stage all the wells show this tendency and the prediction of their behavior in some cases becomes indefinite (stage t_1 in Fig. 2). When the transitory effects have been overcome, the tendency of the well behavior is already defined. In this period the predictions about its future behavior can be made with certainty (stage t_2 in Fig. 2). Later on the behavior of the well will be conserved until reaching its economic limit of lifespan (stage t_3 in Fig. 2).

Weller (1966), beginning with the methodology proposed by Gilbert (1954), established a method by which to estimate the tendencies of decline in a reservoir, through the pressure behavior at bottom-hole well conditions as a function of its production.

In accordance with the above-mentioned method, one can observe (Figs. 1 and 2) that the inflow curves of a well vary with time and are influenced by the reservoir decline.

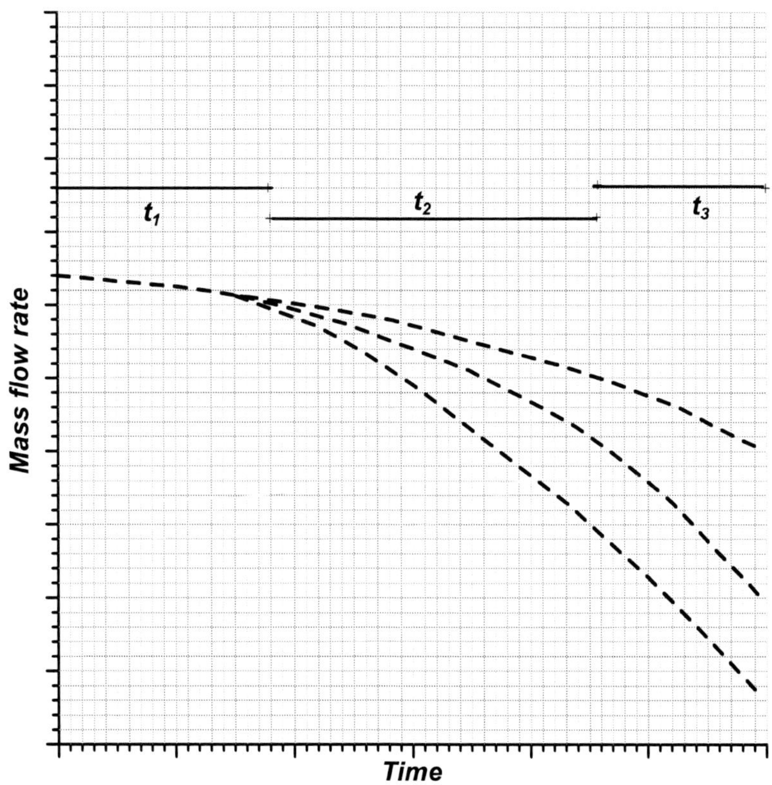

Figure 2. Schematic representation of decline in a well due to its exploitation.

The Inflow Curves for Oil Reservoirs

These types of curves began to be used at the beginning of hydrocarbon exploitation, in order to establish useful approaches in the exploitation designs (Evinger and Muskat, 1942; Muskat, 1945; Gilbert, 1954; Grant et al., 1982). The methods proposed by Fetkovich (1973), Jones (et al., 1976), Chu (1988) and Helmy and Wattenbarger (1998) were applied to practical field case studies.

The techniques applied for this type of analysis were adapted from the results of the pressure transitory analysis (Muskat, 1945; Gilbert, 1954; Van Everdingen and Hurst, 1949; Horner, 1951; Ramey, 1970; Chu et al., 1980).

Weller (1966) established a method to calculate the behavior of the reservoir decline by means of the pressure behavior in the well bottom-hole as a function of production. The above-mentioned technique comprises the determination of well productivity and the implementation of the methodology proposed by Muskat (1945) and Gilbert (1954).

The development, analysis and application of the first relationships of theoretical curves of the inflow behavior, known as “Inflow Performance Relationships” or “IPR”, were made by Vogel (1968). Later on, Standing (1970), Fetkovich (1973), Klins and Majcher (1992), Klins and Clark (1993) and Wiggins (1994), made improvements to these first inflow curves. Basically the different curves were obtained from the following general methodology:

- The construction of the IPR curve at the reservoir average pressure, using the established data of a production test.
- The prediction of the relationship of the inflow behavior, as a function of the reservoir average pressures at different times.

Vogel (1968) made an important innovation to the method of Weller (1966), introducing dimensionless terms in a similar way as the method outlined by Van Everdingen and Hurst (1949) and Horner (1951) for the pressure transitory analyses. The dimensionless parameters that Vogel (1968) uses are dimensionless pressure p_D and dimensionless production rate Q_D.

$$P_D = \frac{P_{wf}}{P_e} \tag{1}$$

$$Q_D = \frac{Q_o}{(Q_o)_{\max}} \tag{2}$$

Where Q_o is the volumetric rate of production determined for a bottom pressure p_{wf}; the static pressure of the reservoir in the feeding area is p_e, the maximum volumetric rate is $(Q_o)_{max}$.

The investigation done by Vogel (1968) included data from wells in 21 different reservoirs. The main result was the proposal of a dimensionless expression known as “Vogel’s equation” or “reference curve of Vogel”:

$$\frac{Q_o}{(Q_o)_{\max}} = 1.0 - 0.2\left(\frac{p_{wf}}{p_e}\right) - 0.8\left(\frac{p_{wf}}{p_e}\right)^2 \tag{3}$$

The variables are the same previously defined.

Standing (1970) essentially extended the application of Vogel's equation to predict the behavior of the inflow, introducing the productivity index. The proposed expression is:

$$Q_o = \left[\frac{J_f^*(p_{e_f})}{1.8}\right]\left[1 - 0.2\left(\frac{p_{wf}}{p_e}\right) - 0.8\left(\frac{p_{wf}}{p_e}\right)^2\right] \tag{4}$$

where J^*_f is the future productivity index, measured at conditions of future pressure in the reservoir p_{ef}.

The proposal of Fetkovich (1973) is given to understand the behavior of wells with turbulent flow; for which he introduced a turbulence factor (τ) in his equation:

$$Q_o = F(p_e^2 - p_{wf}^2)^\tau \tag{5}$$

Where F is an auxiliary variable of this analysis. The values of τ fluctuate between 1.0 (for totally laminar flow) and 0.5 (for highly turbulent flow). As long as there are two unknowns in Eq. (5), then there are required at least two production tests to evaluate both parameters, assuming, also, that the reservoir pressure is known.

Taking the logarithm of both sides of Eq. (5) and clearing log $[(p_e)^2 - (p_{wf})^2)]$, the expression can be written:

$$\log(p_e^2 - p_{wf}^2) = \frac{1}{\tau}\log Q_o - \frac{1}{\tau}\log F \tag{6}$$

Plotting $[(p_e)^2 - (p_{wf})^2)]$ against Q_o in a double logarithmic scale, a straight line is obtained with a slope equal to $1/\tau$ and an interception of F when $[(p_e)^2 - (p_{wf})^2] = 1$. In this way one can determine the value of F at any point of the straight line, after having determined the value of τ.

The methodology of Fetkovich establishes that having determined the values of τ and F it is possible to use Eq. (5) to generate the complete inflow curve of the well.

Klins and Majcher (1992) considered production tests from about 1400 wells, and adjusted the coefficients of the polynomial of second grade of Vogel's equation in the following form:

$$\frac{Q_o}{(Q_o)_{\max}} = 1.0 - 0.1225\left(\frac{p_{wf}}{p_e}\right) - 0.8775\left(\frac{p_{wf}}{p_e}\right)^2 \tag{7}$$

Klins and Clark (1993) improved the predictive capability of Vogel's equation introducing an exponent n denominated as the decline factor. The expression is:

$$\frac{Q_o}{(Q_o)_{\max}} = 1 - 0.295\left(\frac{p_{wf}}{p_e}\right) - 0.705\left(\frac{p_{wf}}{p_e}\right)^n \tag{8}$$

Where:

$$n = \left[0.28 + 0.72\left(\frac{p_e}{p_b}\right)\right](1.24 + 0.001 p_b) \tag{9}$$

In this expression p_b is the boiling pressure of the fluid in the reservoir.
These authors proposed the following procedure to obtain the inflow curves:

- Starting with the analyses of the chemical composition of the fluid, determine the pressure of the boiling point of the fluid and the reservoir pressure.
- Determine the exponent n using Eq.(9).
- Starting with a couple of data (Q_o, p_{wf}) solve Eq. (8) to obtain $(Q_o)_{max}$.
- Build the inflow curve assuming different values of Q_o (from zero until $(Q_o)_{max}$) in Eq. (7).

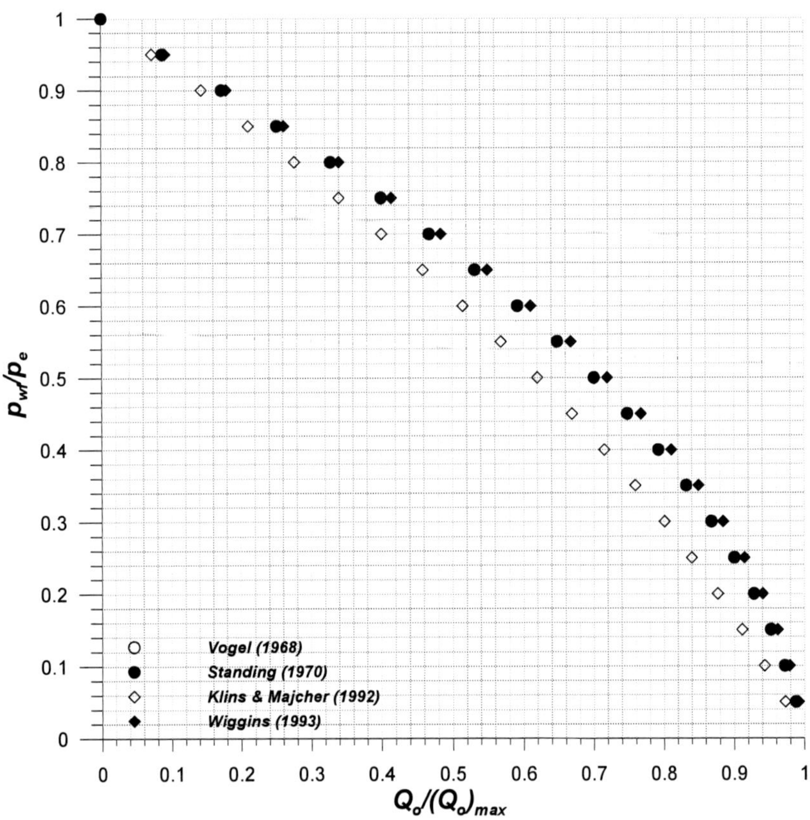

Figure 3. Graphic comparisons of different dimensionless inflow curves proposed by different authors.

Wiggins (1994) proposes a similar Vogel relationship whose form is:

$$\frac{Q_0}{(Q_o)_{\max}} = \left[1 - 0.52\left(\frac{p_{wf}}{p_e}\right) - 0.48\left(\frac{p_{wf}}{p_e}\right)^2\right] \quad (10)$$

Furthermore he also proposed an inflow relationship for water, whose expression is:

$$\frac{Q_{ag}}{(Q_{ag})_{\max}} = \left[1 - 0.72\left(\frac{p_{wf}}{p_e}\right) - 0.28\left(\frac{p_{wf}}{p_e}\right)^2\right] \quad (11)$$

Figure 3 presents the graphical results obtained from the application of Eqs. (3), (4) (7) and (10) corresponding to the relationships of Vogel, Standing, Klins and Majcher and Wiggins respectively. One can observe that the curves obtained with the methods of Vogel, Standing and Wiggins have similar behaviors.

The Damage Effect on the Inflow Curves for Oil Reservoirs

Consistent with previous paragraphs, neither Weller (1966) neither the modifications made by the previously mentioned authors, included in their formulations the effects of the damage effect. Even though the damage effect had already been considered in the analyses of the transitory pressure tests by Warren and Root (1963), Matthews and Russell (1967), Agarwal (et al., 1970), Ramey (1970), Earlougher (1977), Rivera and Ramey (1977), Barua and Horne (1997), Chen and Chang (2006), among other authors.

Klins and Clark (1993) were the first authors to investigate the damage effect on the inflow relationships by incorporating a coefficient M in Eqs. (3), (7) and (8) of Vogel (1968), of Klins and Majcher (1992) and of Klins and Clark (1993) respectively. The resulting expressions are:

$$\frac{Q_o}{(Q_o)_{\max}} = M_1\left[1.0 - 0.2\left(\frac{p_{wf}}{p_e}\right) - 0.8\left(\frac{p_{wf}}{p_e}\right)^2\right] \quad (12)$$

$$\frac{Q_o}{(Q_o)_{\max}} = M_2\left[1.0 - 0.1225\left(\frac{p_{wf}}{p_e}\right) - 0.8775\left(\frac{p_{wf}}{p_e}\right)^2\right] \quad (13)$$

$$\frac{Q_o}{(Q_o)_{\max}} = M_3\left[1.0 - 0.295\left(\frac{p_{wf}}{p_e}\right) - 0.705\left(\frac{p_{wf}}{p_e}\right)^n\right] \quad (14)$$

Variable *M* involves the damage effect s, defined as the relationship between the radius of the reservoir drain area r_e, and the radius of the well r_w. This definition is slightly different for each case and in agreement with Eqs. (12), (13) and (14) the corresponding expressions are obtained for *M*.

$$M_1 = \left(\frac{\ln \frac{r_e}{r_w} - 0.492}{\ln \frac{r_e}{r_w} - 0.492 + s} \right) \tag{15}$$

$$M_2 = \left(\frac{\ln \frac{r_e}{r_w} - 0.476}{\ln \frac{r_e}{r_w} - 0.476 + s} \right) \tag{16}$$

$$M_3 = \left(\frac{\ln \frac{r_e}{r_w} - 0.467}{\ln \frac{r_e}{r_w} - 0.467 + s} \right) \tag{17}$$

Variable *M* influences the rate in Darcy's Law in an inverse way. The values of the coefficients (0.492, 0.476 and 0.467) were obtained from the simulations made by Klins and Majcher (1992), consonant with the values of the coefficients of p_{wf}/p_e of Eqs. (12), (13) and (14) respectively. The values of these coefficients are close to 0.5 and the origin of the variation is the result of the values of the different coefficients used in the simulations.

Considering the typical values of r_e and of r_w for oil systems, the previous expressions of *M* are simplified to:

$$M_1 = \left(\frac{6.81}{6.81 + s} \right) \tag{18}$$

$$M_2 = \left(\frac{6.886}{6.886 + s} \right) \tag{19}$$

$$M_3 = \left(\frac{6.835}{6.835 + s} \right) \tag{20}$$

In Eq. (8) the exponent *n* is the decline factor, Eq. (9), which is a function of the reservoir pressure (p_e) and the boiling pressure (p_b).

The variation of n becomes important when the boiling pressure p_b is much smaller than the reservoir pressure p_e; which implies a prominent two-phase flow. Figure 4 shows the behavior of Eq. (8) for reservoir pressures between 206.8 and 620.4 bar (3000 and 9000 psia approximately) and boiling pressures between 55.1 and 275.7 bar (800 and 4000 psia approximately). The value of n is found within an interval of 4 and 16. This value is modified during the operative life span of the well, it being a function of both the reservoir pressure and the boiling pressure.

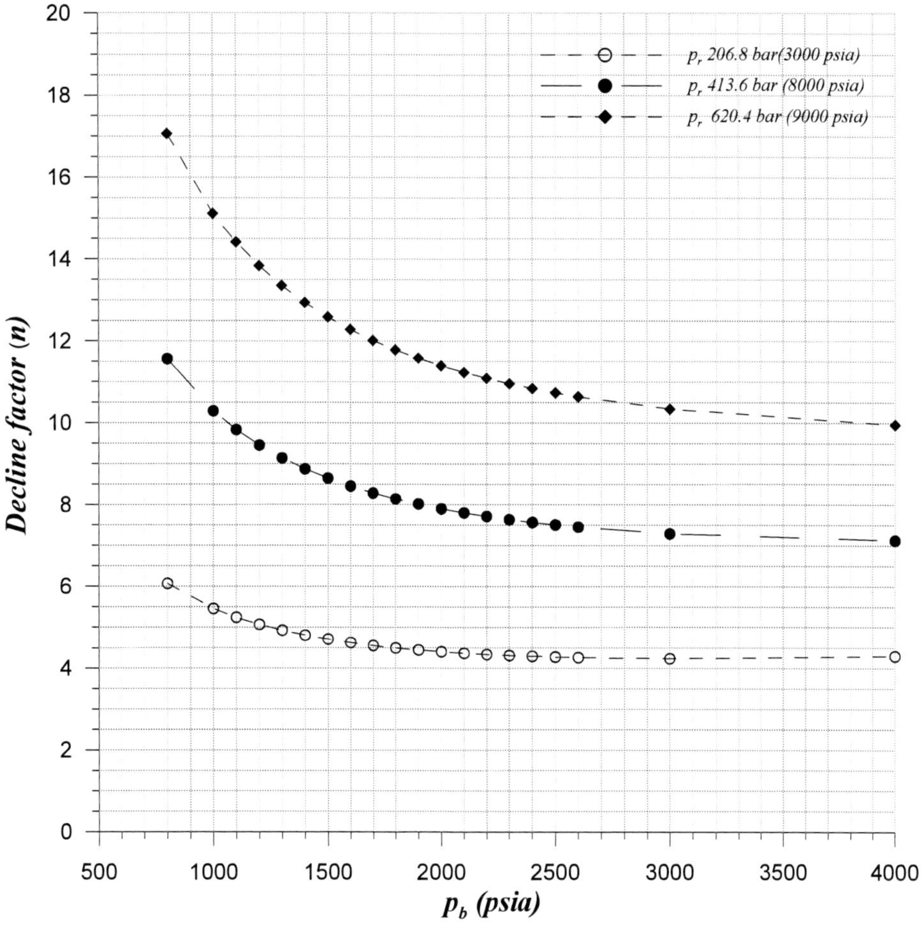

Figure 4. Intervals of the decline factor at average conditions of an oil system.

In Fig. 5 the graphics of Eq. (14) proposed by Klins and Clark are shown, for different values of n, within the interval of 4 and 14.

It is possible to observe that the change in the value of n depends mainly on the variation of the reservoir pressure (p_e), which diminishes faster with respect to the change of the boiling pressure (p_b), (Standing, 1970). As previously mentioned, the variation of the decline factor n influences the inflow curves of the well, just as it modifies its productivity.

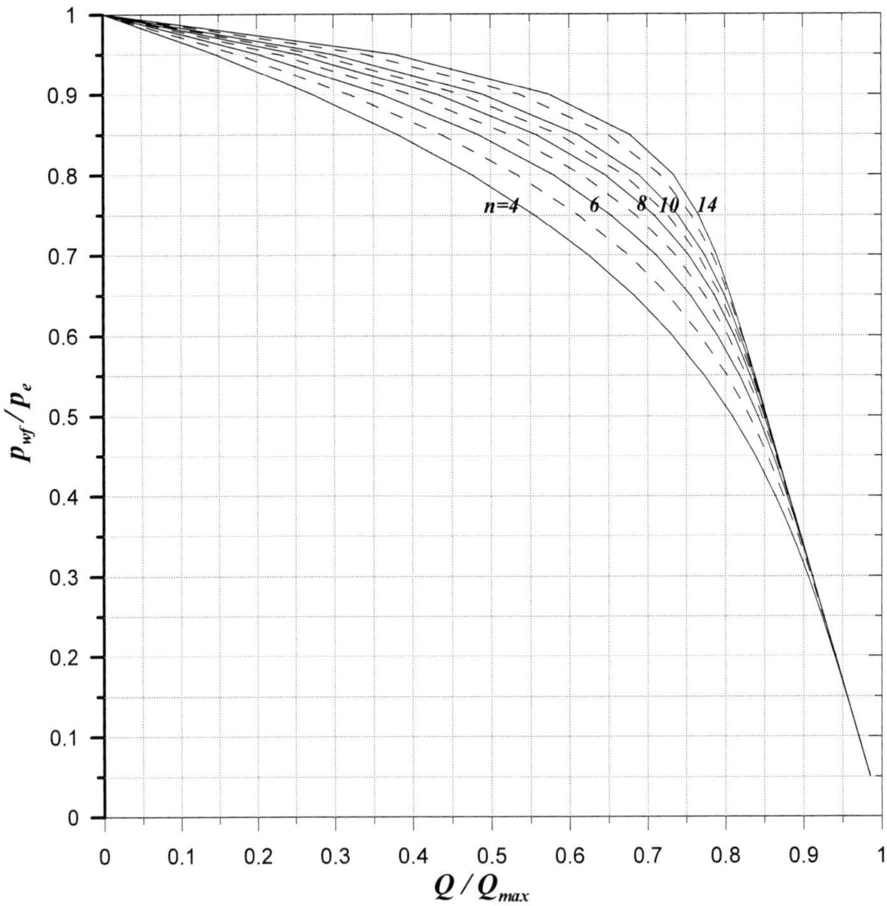

Figure 5. Sensibility of the reference curve (Eq. 14) of Klins and Clark (1993), with respect to the variation of the decline factor *n*.

Inflow Type-Curves with Damage Effect for Oil Reservoirs

Figure 6 compares the curves for the damage values of -4, 0 and 4 respectively. These curves were calculated from the inflow relationships of Vogel (1968), of Klins and Majcher (1992) without considering decline, and of Klins and Clark (1993) considering decline.

Of all the previously presented inflow relationships the most complete is Eq. (14), because it includes simultaneously the damage effect in the well and its decline. This one is presently used in different productivity diagnoses and to estimate the damage in oil wells (Gallice and Wiggins, 2004; Al Qahtani, 2001).

Figure 7 shows the graphics of the dimensionless inflow curves calculated from the relations proposed by Klins and Clark (1993), Eq. (14) that includes decline.

The relevance of inflow relations in reservoir engineering and production of hydrocarbons resides in its application to the diagnosis of productivity of the wells. Even more, this diagnosis becomes complete when it considers the damage effect on the productivity.

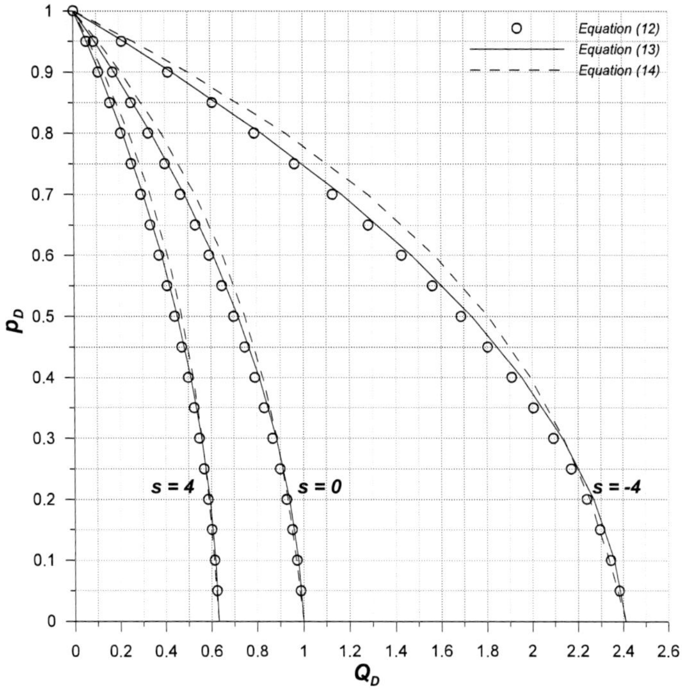

Figure 6. Dimensionless inflow curves for skin factors of -4, 0 and 4. They were calculated using the inflow relations of Vogel (1968), Klins and Majcher (1992), without considering decline and Klins y Clark (1993), with decline ($n = 2.17$).

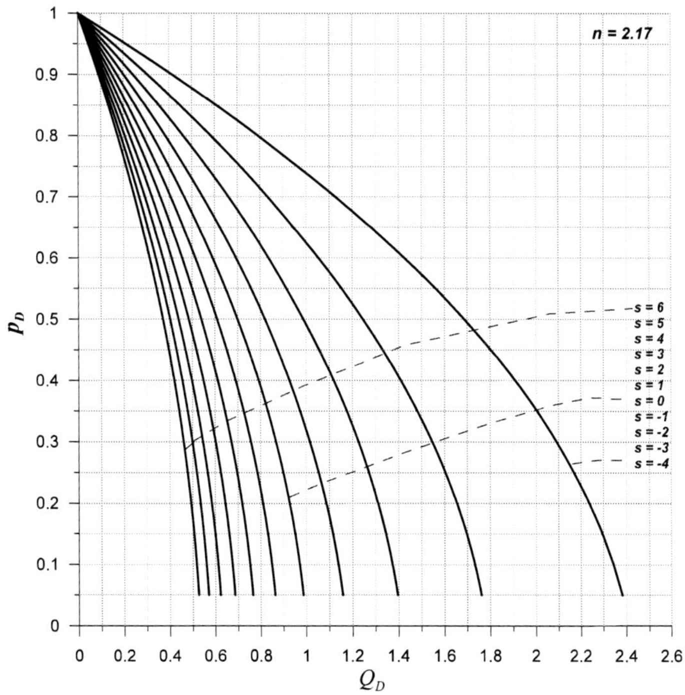

Figure 7. Inflow dimensionless curves for different damage values, calculated using the inflow relation of Klins and Clark (1993) affected by the damage effect, and considering a decline of 2.17.

Inflow Curves for Geothermal Reservoirs

In a geothermal reservoir the main resource is its thermal power. This power is extracted through the produced fluid. The production of the fluid is influenced by pressure, temperature, enthalpy, reservoir dimensions, permeability and by the thermo-physical properties of the reservoir.

In reservoir geothermal engineering, correlations and methodologies from oil reservoirs are frequently applied to the analysis of the behavior of this type of system. In this section a revision is made of the different inflow and outflow curves obtained for geothermal systems.

James (1968; 1970; 1980; 1983; 1989), Goyal et al., (1980) and Grant et al., (1982), among others, found that the outflow curves fashioned a solid tool for the analysis of wells and reservoir characterization. Garg and Kassoy (1981) analyzed different outflow curves to determine the thermal power of different wells. Kjaran and Elliasson (1982) used data from the Icelandic geothermal field in Svartsengi and obtained a general equation of outflow curves representative of the whole field.

The particular concavity of the outflow curves of each well is discussed by Grant et al., (1982) who described the form of each curve with respect to the different behavior of these wells. In accordance to this approach, they were able to identify wells fed by reservoir water; wells located in high permeability areas; vapor dominated wells or those with gas content, along with wells producing two-phase fluid.

Other parameters, besides pressure and mass flow, associated with outflow curves are enthalpy, temperature, density and viscosity. Chu (1988) uses and compares the techniques outlined by Fetkovich (1973) and Jones et al., (1976) finding applicability in the diagnosis of well conditions. Conversely, Chu also demonstrated that the productivity index of geothermal wells evolves as a function of two reservoir parameters: the extracted fluid volume and the period of exploitation. Leaver and Freeston (1987) apply the method of Ramey (1981) to predict outflow curves in steam wells.

Iglesias and Moya (1990) formulated the first dimensionless inflow curve for geothermal reservoirs, considering the geothermal fluid as only pure water. The expression is:

$$\frac{W}{W_{\max}} = 1.0 - 0.6\left(\frac{p_{wf}}{p_e}\right)^2 - 0.4\left(\frac{p_{wf}}{p_e}\right)^4 \tag{21}$$

where W is the produced mass flow, W_{max} is the maximum mass flow (theoretically for $p_{wf} = 0$). The flowing bottom pressure is p_{wf} and the reservoir pressure is p_e.

The same authors presented the corresponding inflow curve for thermal productivity (thermal power):

$$\left(\frac{Pot}{Pot_{\max}}\right) = 0.7\left(\frac{W}{W_{\max}}\right) + 0.3\left(\frac{W}{W_{\max}}\right)^2 \tag{22}$$

where *Pot* is the *thermal power* = (*W*) (*H*), and *H* being flowing enthalpy.

Subsequently, Moya (1994) obtained the respective dimensionless inflow curves for a binary system H_2O-CO_2, being the expression of the mass productivity as follows:

$$\frac{W}{W_{max}} = 1.0 - 0.256\left(\frac{p_{wf}}{p_e}\right) - 0.525\left(\frac{p_{wf}}{p_e}\right)^2 - 0.057\left(\frac{p_{wf}}{p_e}\right)^3 - 0.162\left(\frac{p_{wf}}{p_e}\right)^4 \quad (23)$$

Moya et al. (1995, 1997, 1998) made applications of the binary model to field case studies of Mexican geothermal reservoirs. They estimated outflow curves and compared them to field data. Iglesias and Moya (1998) validated the inflow curves comparing their results with bottom-hole well data. Moya et al. (2001, 2003) extended the application of this methodology to estimate the permeability of rock formations by means of a computation system (Moya and Uribe, 2000) that applies the methodology in automated form.

To introduce the effect of salts, Montoya (2003) proposed an inflow curve that considers the geothermal fluid to be a ternary mixture H_2O-CO_2-$NaCl$. This expression assumes low salt content (up to 5% of mass fraction in the liquid phase) and its form is:

$$\frac{W}{W_{max}} = 0.999 - 0.436\left(\frac{p_{wf}}{p_e}\right) - 0.537\left(\frac{p_{wf}}{p_e}\right)^2 + 0.694\left(\frac{p_{wf}}{p_e}\right)^3 - 0.715\left(\frac{p_{wf}}{p_e}\right)^4 \quad (24)$$

For high salt content (30% of mass fraction in the liquid phase) including precipitation conditions, Meza (2005) proposes the following expression:

$$\begin{aligned}\frac{P_{wf}}{P_e} &= 1.0 - 0.619\left(\frac{W}{W_{max}}\right) + 1.45\left(\frac{W}{W_{max}}\right)^2 \\ &= -5.476\left(\frac{W}{W_{max}}\right)^3 + 7.605\left(\frac{W}{W_{max}}\right)^4 - 3.955\left(\frac{W}{W_{max}}\right)^5\end{aligned} \quad (25)$$

The same expression can be written as:

$$\begin{aligned}\frac{W}{W_{max}} &= 1.0 - 0.4399\left(\frac{P_{wf}}{P_e}\right) + 1.1658\left(\frac{P_{wf}}{P_e}\right)^2 \\ &= -4.0372\left(\frac{P_{wf}}{P_e}\right)^3 + 3.6697\left(\frac{P_{wf}}{P_e}\right)^4 - 1.3782\left(\frac{P_{wf}}{P_e}\right)^5\end{aligned} \quad (26)$$

Figure 8 shows the comparison among these four inflow curves for geothermal reservoirs corresponding to Eqs. (21), (23), (24) and (26).

The curve for pure water was obtained using the geothermal reservoir simulator SHAFT79 (Pruess and Shroeder, 1980). For the binary mixture and for the ternary mixture, at low salt concentrations, the simulator BIOX (Iglesias and Moya, 1990; 1995) was used.

BIOX includes a subroutine to compute the CO_2 solubility in water. For high salinity relationship the TOUGH2 simulator was used (Pruess et al., 1999).

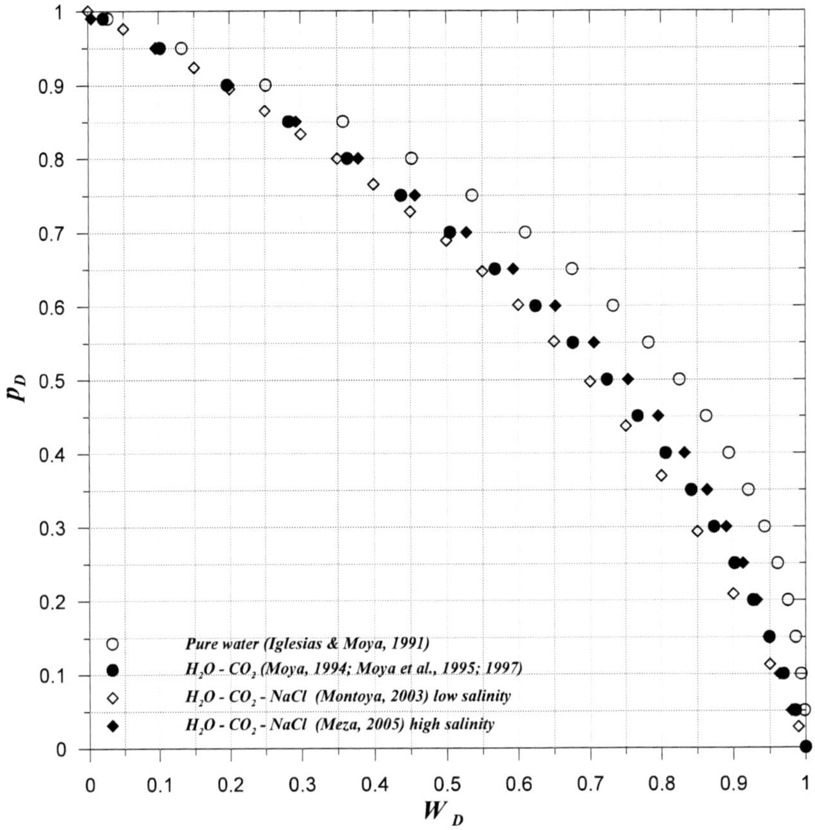

Figure 8. Proposed dimensionless inflow curves of mass productivity for geothermal systems.

Contrary to the practice in petroleum engineering, dimensionless inflow curves with damage effect have still not been developed in geothermal engineering.

Dimensionless Inflow Curves with Damage Effect Proposed for Geothermal Reservoirs

To include the damage effect in geothermal inflow curves, it is convenient to take into account the equations governing the reservoir flow and the inflow curves at the interface reservoir-well. Eq. (27) shows that the damage effect is associated with the magnitudes of both the drain radius of the reservoir and the radius of the well.

$$Q_o = \frac{0.00708Kh(p_i - p_{wf})}{\mu_o B_o \left[\ln \frac{r_e}{r_w} + s \right]} \tag{27}$$

Klins and Majcher (1992), along with Klins and Clark (1993), in their proposal of the parameter M (see Eqs. (18), (19) and (20)), they utilized a similar expression to the denominator in Eq. (27). In determining the values of the coefficients in the afore- mentioned equations, Klins and Majcher assumed a reservoir drain radius between 220 and 300 ft and an average radius for the wells of about 2.5 pg. The previous values are characteristic of the oil systems and the final result is the combination of Eqs. (14) and (20). The graphic representation of these equations is shown in Fig. 7, which is the inflow curve for different values of damage, applicable to oil systems.

Similarly for geothermal reservoirs, according to different authors, it is feasible to assume that the average drain radius r_e, is around 400 m (1312 feet) and about 3.5 inches for the well radius r_w, (Grant et al., 1982; Gunn and Freeston, 1991). Therefore, the value of M in a geothermal system is expressed as follows:

$$M = \frac{\ln\frac{r_e}{r_w} - 0.6603}{\ln\frac{r_e}{r_w} - 0.6603 + s}$$

$$M = \frac{7.75}{7.75 + s} \qquad (28)$$

Fundamentally, the damage in a well is manifested as a decrease in its productivity. However the concept of damage includes cases in which the original conditions of the well are modified. Consistent with this, the damage can be positive, null or negative.

Positive damage values indicate decrease in productivity; while negative values indicate improvement in productivity. Zero damage indicates that the productivity is the natural effect of the reservoir without any artificial manipulation. A negative damage value appears in washed wells, stimulated wells or fractured wells improving productive characteristics. In accordance with Eq. (28) the damage factor, damage effect or skin factor (s) is an inverse function of the parameter M. As the value of s increases, the value of M diminishes. Conversely, the more s diminishes, M will increase. M affects inflow relationships in percentage form.

Figure 9 shows the behavior of M as a function of the damage effect s. The behavior of M, regarding the skin factor, helps the reservoir engineer in developing exploitation strategies for the well. Beginning with Fig. 9 one can observe that for damage values bigger than 10, the productivity diminishes to values inferior than 40%. The preceding result leads to the conclusion, that when the well arrives to these limits of productivity, it becomes necessary to make a technique-economic analysis concerning the continuity of the project and its exploitation.

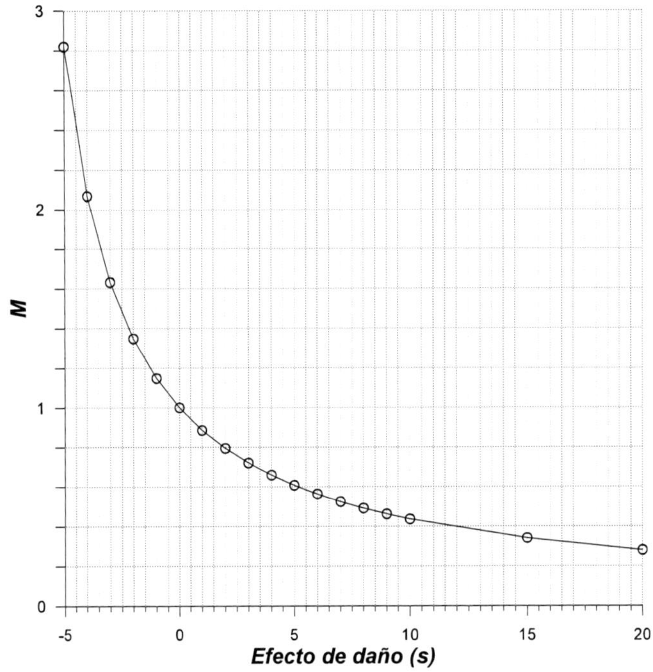

Figure 9. Behavior of parameter M (Eq. 28) as a function of the skin factor s.

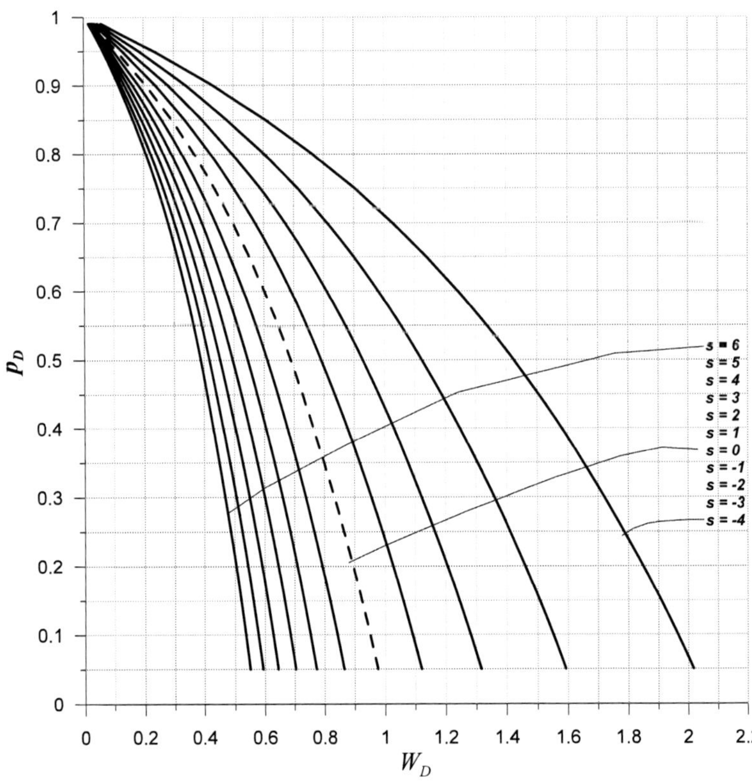

Figure 10. Type-curves for different skin factors s, obtained from the dimensionless inflow curve for geothermal reservoirs with H_2O -CO_2 – $NaCl$ (low salinity mixture).

Incorporating parameter M inside the geothermal inflow relationships considering a fluid constituted by H_2O - $CO_2 - NaCl$ at low (Eq. 24) and high salinity (Eq. 26), one has:

$$\frac{W}{W_{max}} = M\left[0.999 - 0.436\left(\frac{P_{wf}}{P_e}\right) - 0.537\left(\frac{P_{wf}}{P_e}\right)^2 + 0.694\left(\frac{P_{wf}}{P_e}\right)^3 - 0.715\left(\frac{P_{wf}}{P_e}\right)^4\right] \quad (29)$$

$$\frac{W}{W_{max}} = M\left[1.0 - 0.4399\left(\frac{P_{wf}}{P_e}\right) + 1.1658\left(\frac{P_{wf}}{P_e}\right)^2 - 4.0372\left(\frac{P_{wf}}{P_e}\right)^3 + 3.6697\left(\frac{P_{wf}}{P_e}\right)^4 - 1.3782\left(\frac{P_{wf}}{P_e}\right)^5\right] \quad (30)$$

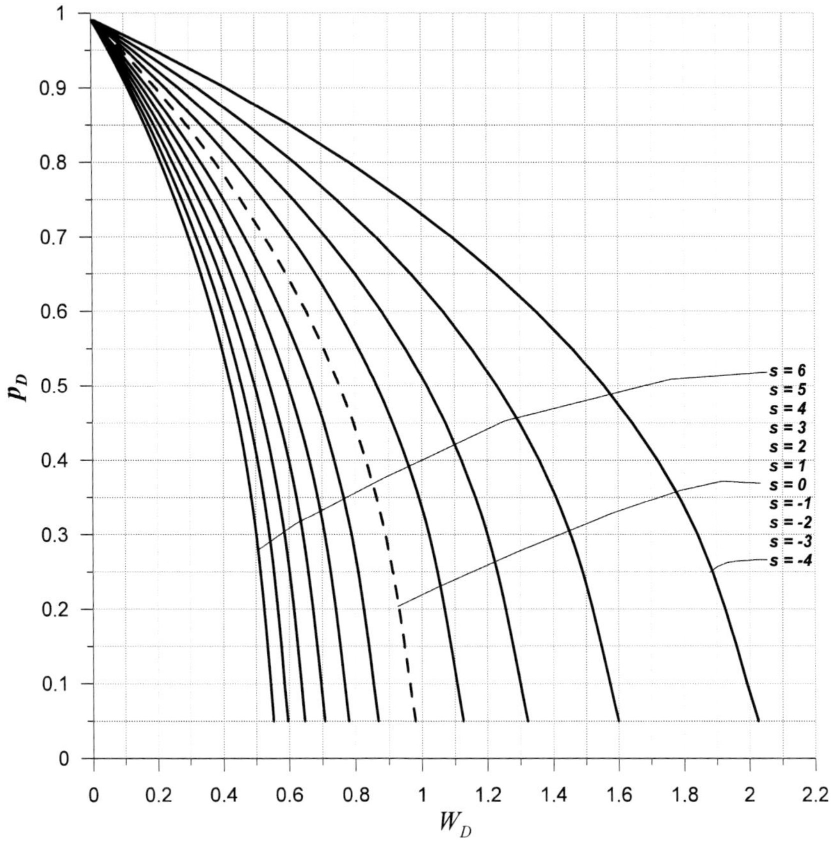

Figure 11. Type-curves for different skin factors s, obtained from the dimensionless inflow curve for geothermal reservoirs with H_2O - CO_2– $NaCl$ (high salinity mixture).

Figs. 10 and 11 represent Eqs. (29) and (30) for different values of damage. Basically they represent the main product of this chapter and are designated as geothermal type-curves with damage effect (for low and high salinity, respectively). Their utility rests in the determination of the damage value in wells starting with the dimensionless values of inflow from their production tests.

The established approach for detecting a low salinity mixture is considering the salt content of its mass fraction, up to 5%. A high salinity mixture is identified by a mass fraction superior to 5% (Pruess et al. 1999).

Proposed Methodolgy to Determine the Skin Factor

The proposed methodology in determining the damage in wells by using geothermal type-curves with damage effect [Eqs. (29) and (30)] is shown in the flow diagram in Fig. 12.

In practical terms the process involves the following points:

1. The input data correspond to a production test: mass flow (W), flowing pressure and enthalpy (p and H), and reservoir static pressure (p_e).
2. If the available data are obtained at wellhead conditions, a well simulator is used to obtain the bottom-hole conditions.
3. The dimensionless inflow curve of the well is determined from the reference curves [Eqs. (29) and (30)] using the computational system SISTCURV (Moya and Uribe, 2000).
4. Overlapping the dimensionless curve of the well with the geothermal type-curves with damage effect (Figs. 10 or 11 according to the case of low or high salinity). The value of s for the best overlapping is the optimal value.

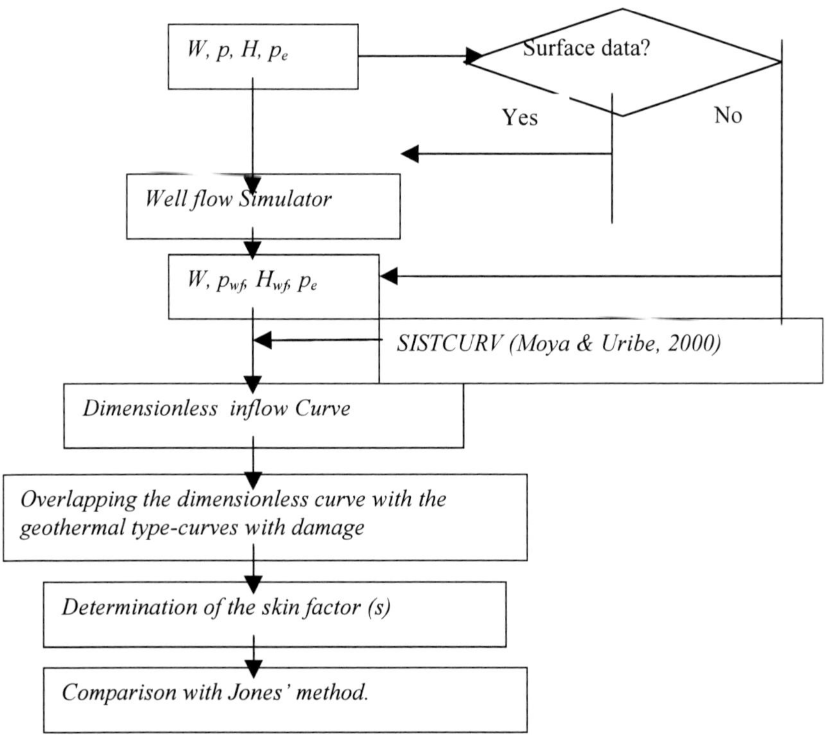

Figure 12. Methodology to determine the skin factor in a geothermal well, using type-curves with damage effect.

Since the proposed methodology is new and there is no way of verifying the results, we investigated and found that Jones' method allows the identification of the presence or absence of damage, thereby adopting said methodology for this work in order to check the damage values of the production tests analyzed. The application of the proposed methodology in combination with Jones' method was successful in the development of this work. The diagnosis obtained with this technique was congruent in all cases with the damage value estimated with the type-curves and proposed methodology. In accordance with these results, the Jones' method is described next.

Revision of Method Used by Jones, Blount and Glaze (1976)

The Jones' (et al., 1976) method is used to determine formation conditions at the end of the perforation as well as at any stage in the life of the well. It was designed primarily for application in oil wells, but, to date, such application has not been extended to geothermal wells. This may be due to the fact that when the wells are damaged, the diagnosis of their mechanical condition is made through contracts with the companies in charge of carrying out the repair and/or the fracturing. The method is useful in the identification of pressure losses for turbulent flow (there are restrictions in the well's feeding area related to the damage).

The method requires at least three pairs of data (mass rate, pressure) measured during a production test. The procedure consists of the following points:

Calculate $(p_e\text{-}p_{wf})/q$ for the different measured mass rates. This is equivalent to obtaining the inverse of the productivity index, because:

$$J = \frac{q}{\Delta p} = \frac{q}{p_e - p_{wf}} \tag{31}$$

where J is the productivity index.

- Make a plot of $(p_e\text{-}p_{wf})/q$ vs. q
- Adjust the calculated points to a straight line and obtain its equation determining the ordinate to the origin b, and the slope m.

$$(p_e\text{-}p_{wf})/q = mq + b \tag{32}$$

For the diagnosis it's necessary to calculate b', whose expression is:

$$b' = b + mQ_{max} \tag{33}$$

Where Q_{max} is obtained from Eqs. (24) or (26) at $p_{wf} = 0$ conditions. On the other hand for a given mass rate one can determine the reservoir pressure applying the expression:

$$p_e - p_{wf} = bQ + mQ^2 \tag{34}$$

It is convenient to emphasize that the method is designed to use the pressures in psia and the flow in *brl / d.*

The criteria of Jones' method applied in the diagnosis are:

- If the value of the ordinate b is smaller than 0.05 the formation is not damaged. Consequently for values bigger than 0.05, there is the presence of damage.
- For values of $b' /b < 2.0$ the turbulence is small or null at the interface well-reservoir.
- For small b and high values of b' /b (without damage and without turbulent flow) the low productivity is caused by the insufficiency of the available flow area. In this case the appropriate recommendation is to enlarge the exploitation zone by deepening the well. In oil wells, low productivity can be corrected by simply enlarging the area or increasing the density of shots in the production pipe.
- When b is larger than 0.05 and b' is small, there is damage and the productivity is not good, therefore well treatment is recommended. This treatment could be a simple cleaning, or stimulation and/or fracturing.

In petroleum reservoir engineering, Jones' method is generally used in the diagnosis of well conditions. Fundamentally this method is used when the wells show decline in their productivity. However even when the application of Jones' method in geothermal reservoir engineering is not very common it was found that, in order to verify the results of this research, the obtained diagnoses are linked to the determination of the damage value using the proposed methodology.

From the previous diagnosis criteria, the first is the one used in this work to confirm the presence or absence of damage. It is necessary to clarify that Jones' method is limited to only diagnosing the presence or absence of damage; it is not used to determine its value. Therefore it is just a qualitative method. The identification of the presence or absence of damage is enough to corroborate the veracity of the results obtained using the proposed methodology. A negative value means absence of damage; a positive value means damage.

Examples of Application of This Methodology to Determine the Damage Effect Using Geothermal Inflow Type-Curves

The inflow type-curves utilized assume that the fluid is a compound H_2O-CO_2-$NaCl$. This composition is the one most similar to a geothermal fluid, because it considers gases and salts. The limits for low salinity are for saline concentrations up to 5000 ppm and those for high salinity are saline concentrations larger than 5000 ppm. Using these considerations the methodology was applied, and in order to show its application, in each case we used type-curves for low and high salinity.

In order to show the validity of the type-curve with damage effect [Eqs. (29) and (30)], data from five production tests were used. These tests are published in the technical world literature. In all cases the previously described methodology was applied.

The Carry City Well

This well, located in Oklahoma (USA), and the average pressure of the reservoir is 110.3 bar (1600 psi). The production test employed took place over approximately a two week period, varying the mass rates of the well in sequential form. Millikan and Sidewell (1931), Gallice and Wiggins (2004) published the measured data at bottom-hole conditions in the production test. The data are shown in Fig. 13. The respective dimensionless data of the well are overlapped to the type-curve with damage effect for low and high salinity. The results are shown in Figs. 14 and 15.

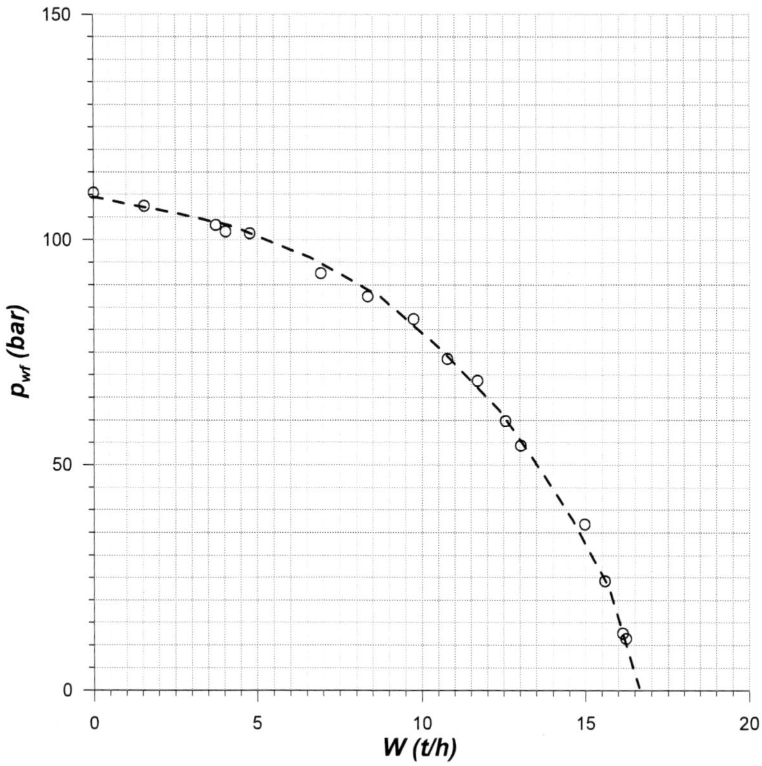

Figure 13. Inflow curve obtained from the data of the production test in the Carry City well of Oklahoma (Milikan and Sidewell, 1931; Gallice and Wiggins, 2004).

Figures 14 and 15 show that the largest quantity of dimensionless well data is located between the curves -1 and 0.9. The negative values are in the area of low flows, for which there is greater uncertainty because more time is required to reach stable conditions. As such, taking into consideration the last five values, the averages are s = 0.4 and s = 0.6 in conformity with, respectively, the type-curves with damage effect for low and high salinity.

The following stage consists in verifying the presence or absence of damage using the Jones' method (Jones et al., 1976). Figure 16 shows the data in the suggested form of the mentioned method.

Upon adjusting the data a straight line is obtained, whose equation is:

$$\Delta p/Q = 0.000187\,(Q) + 0.063$$

The value of the ordinate is $b = 0.063 > 0.05$. Then, in accordance with Jones' method, the resulting diagnosis is that there is damage in the well.

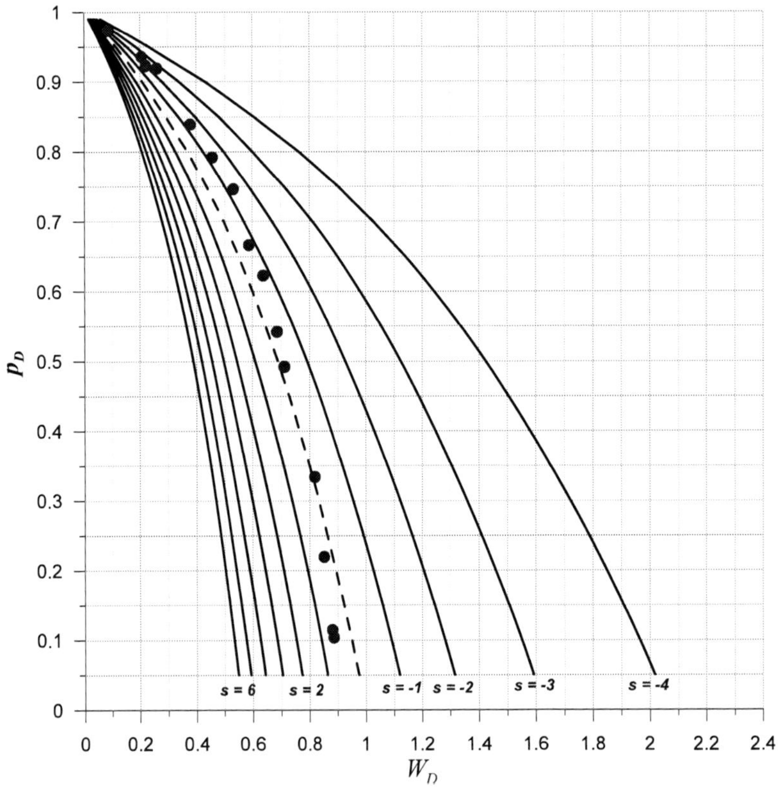

Figure 14. Inflow data from Carry City well overlapped to the type-curves with damage effect for geothermal wells with low salinity.

In the previous stage (Figs. 14 and 15) values of damage for $s = 0.4$ and $s = 0.6$ were obtained, whereby both results show consistency with Jones' method in that they diagnose the presence of damage in the well. The dispersion of field data might recommend, in some cases, the application of regression methods other than linear regression (Verma et al., 2006), in order to identify whether or not the dispersion is a result of the well's response or if it is simply produced by the sensor.

In the following examples of analysis the process will no longer be detailed. Only a description of the analysis will be made and the corresponding plots will be shown. The sequence of the proposed methodology will be conserved.

The corresponding outflow curves will be shown in sequential form when wellhead data are available. The inflow curve, overlapping with the type-curve with damage effect for low and high salinity and the result of the application of Jones' graphic method will be shown.

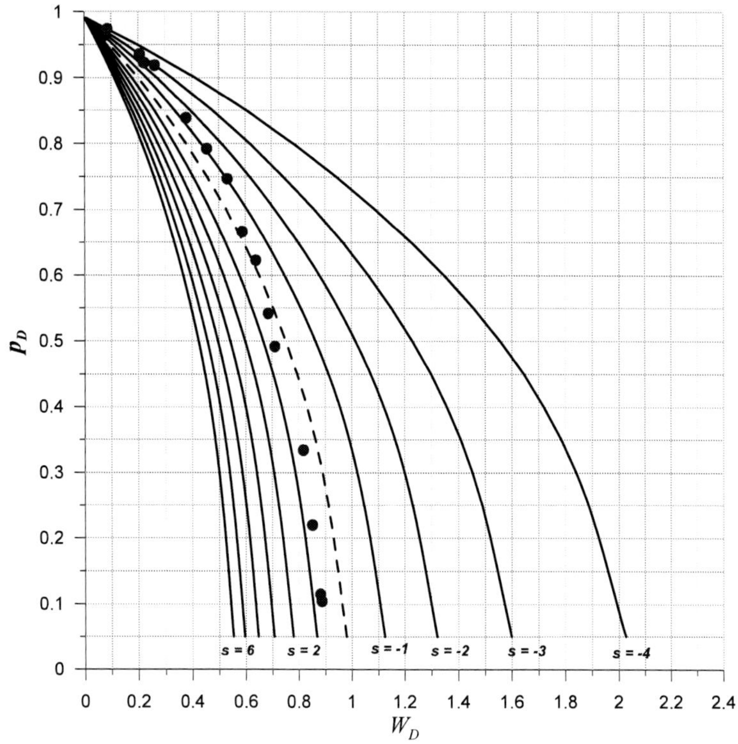

Figure 15. Inflow data from Carry City well overlapped to the type- curves with damage effect for geothermal wells with high salinity.

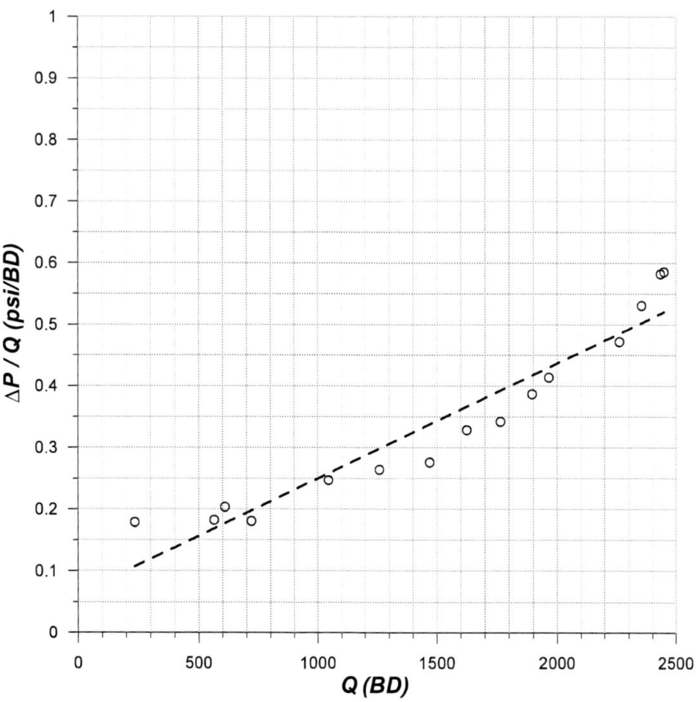

Figure 16. Jones' method applied to the well in Carry City.

The Sweezy-1 Well

This well is located in Jolla, California, USA. In it a production test was made with measurements at well bottom-hole (Garg and Riney, 1984; Min H. Chu, 1988). The method

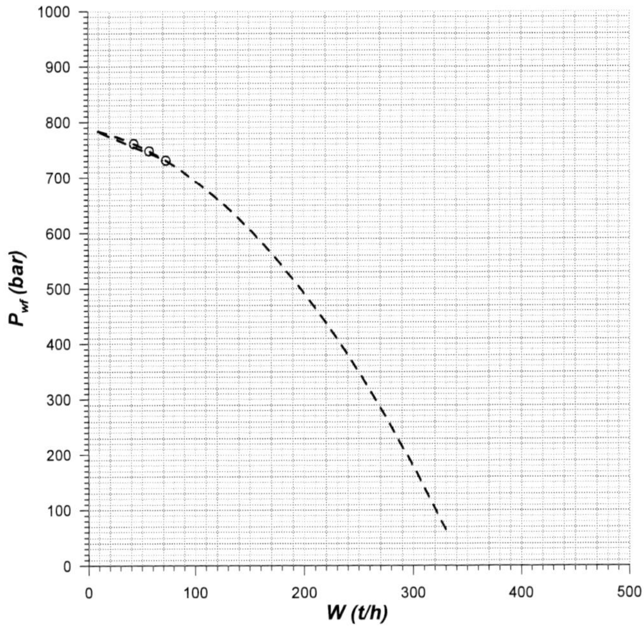

Figure 17. Inflow curve obtained at the Sweezy-1 well (Garg y Riney, 1984; Min H. Chu, 1988).

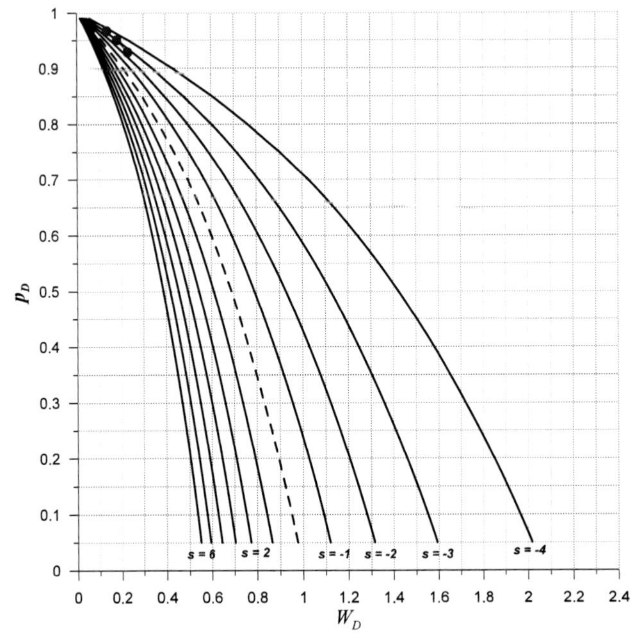

Figure 18 Overlapping of the inflow data of the Sweezy-1 well with the type- curves with damage effect for geothermal wells with low salinity, resulting that $s = -2.5$.

was applied with only three values, due to the fact that the well was analyzed by the original authors and the data are useful in the description of the methodology.

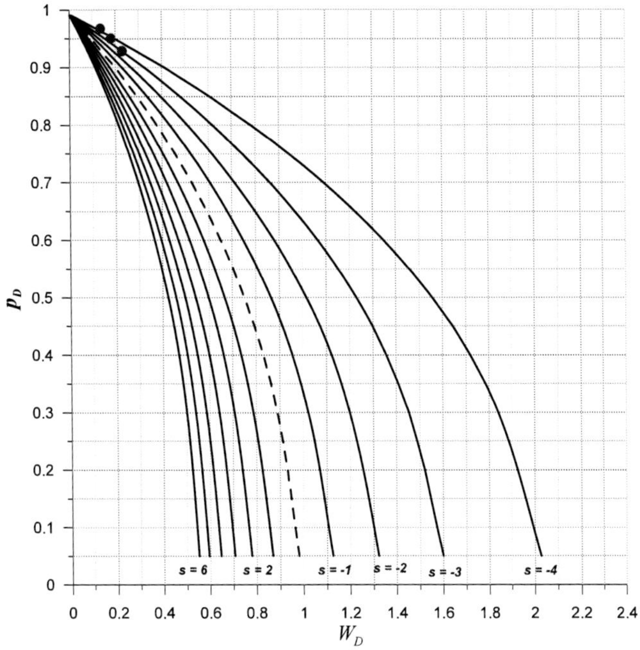

Figure 19. Overlapping of the inflow data of the Sweezy-1 well with the type- curves with damage effect for geothermal wells with low salinity, resulting that s = -2.2.

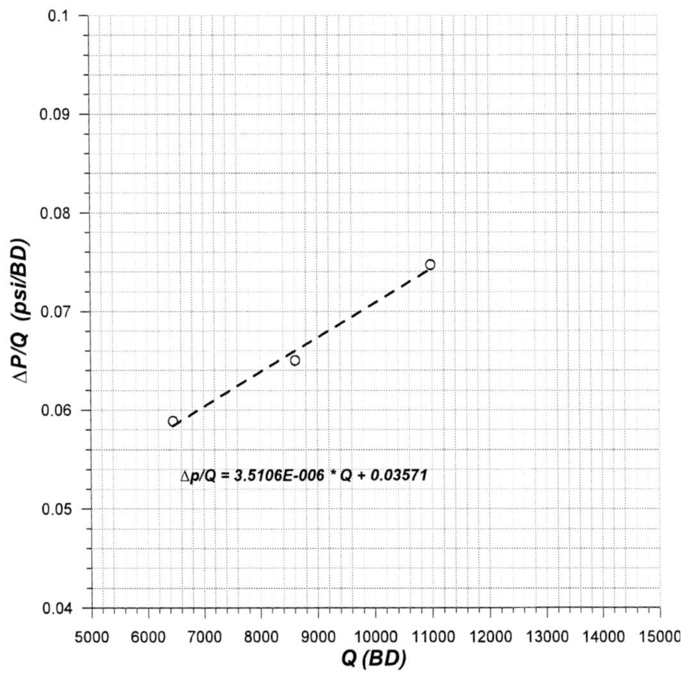

Figure 20. Confirmation of the absence of damage in Sweezy-1 well using Jones' method.

According to Fig. 20, the value of the ordinate is $b = 0.036$, which is smaller than 0.05. Therefore, it is concluded damage does not exist. This agrees with the results obtained from the overlapping of the well's values in the type-curves for low and high salinity.

Well M-110

This well is located in the Cerro Prieto, Mexico geothermal field. The data were published by Ribó (1989) and obtained from two production tests, one performed in 1979, and another one in 1985. Data was obtained from a third production test made in 1987, after a larger operative life of the well.

The tests were made under initial conditions in the exploitation of the well. In the first case after 6 years of exploitation and, in the second case, after 8 years of exploitation.

The characteristic production curves were built with data measured at surface conditions. As shown in Fig. 21, one can observe the decline in the productive characteristics of the well.

The application sequence of the methodology, the determination of the well skin factor and the corresponding confirmation of the results using Jones' method, along with well data, is shown in a single graph for each test throughout its operative life.

The corresponding plots are shown in Figs. 22, 23 and 24 for the well production tests made in 1979, 1985 and 1987.

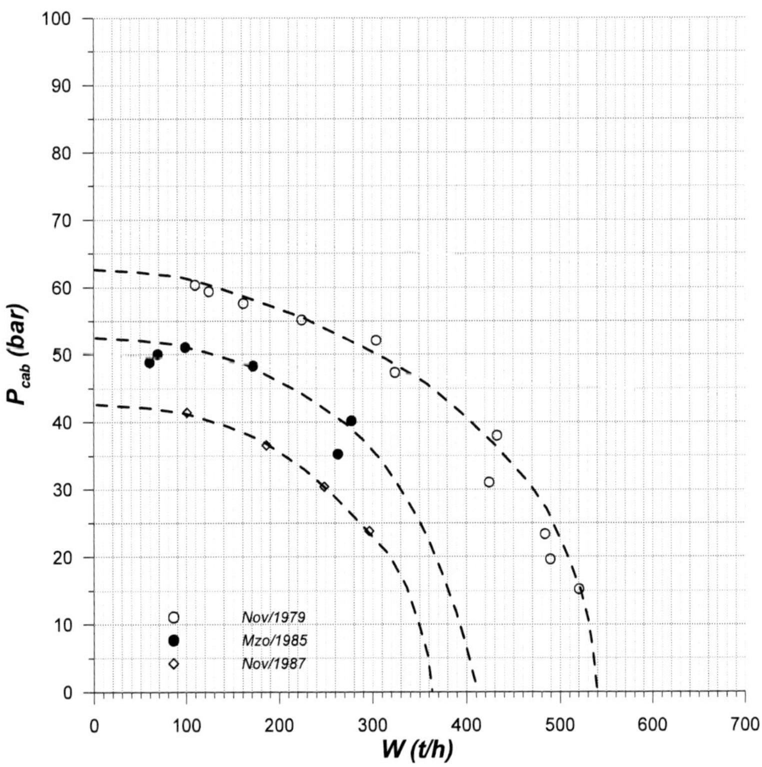

Figure 21. Characteristic production curves of well M-110 obtained from the data published by Ribó (1989).

Well M-110 (Production Test of 1979)

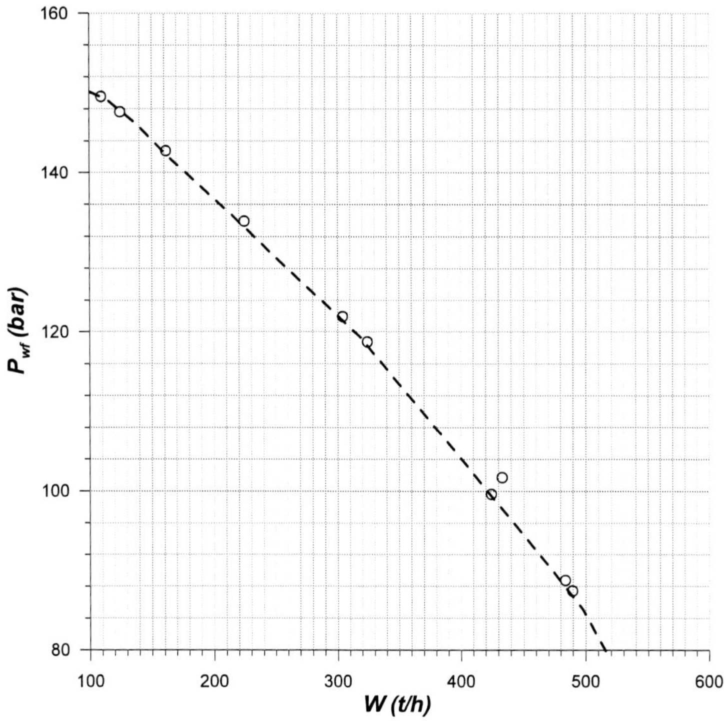

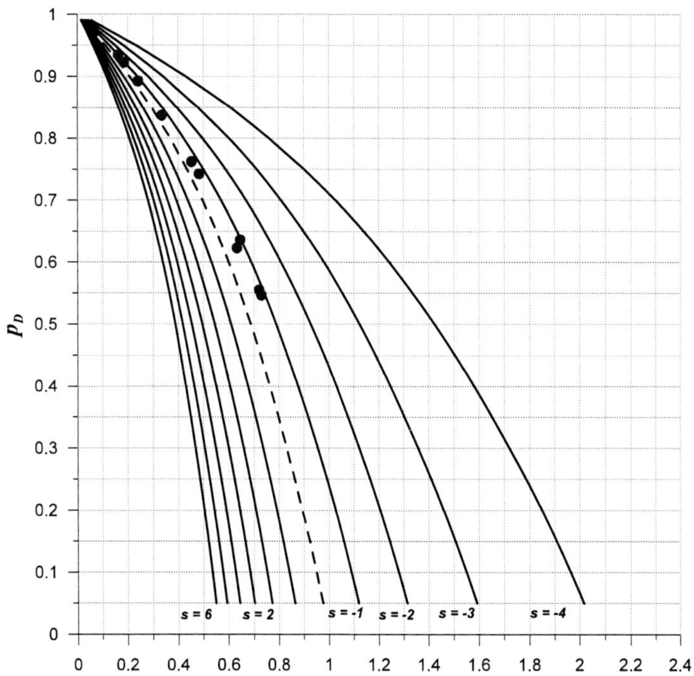

Figure 22. Continued on next page.

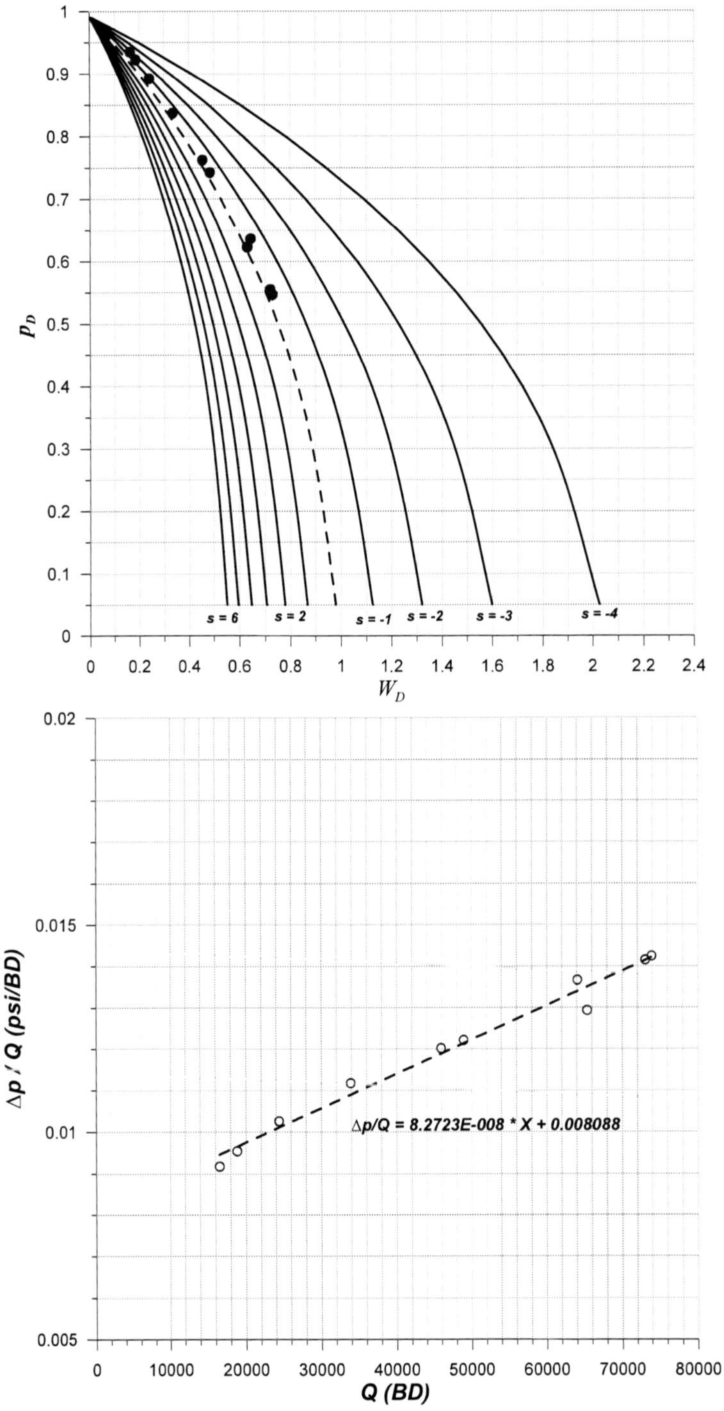

Figure 22. Overlapping of the inflow curve of well M-110 obtained in 1979 with the type- curves for low and high salinity. The diagnosed values of damage s = -0.5 and s = -0.2 imply that there is no damage in the well. The ordinate b = 0.008, - - - calculated with Jones' method, confirms the absence of damage, because b<0.05.

Well M-110 (Production Test of 1985)

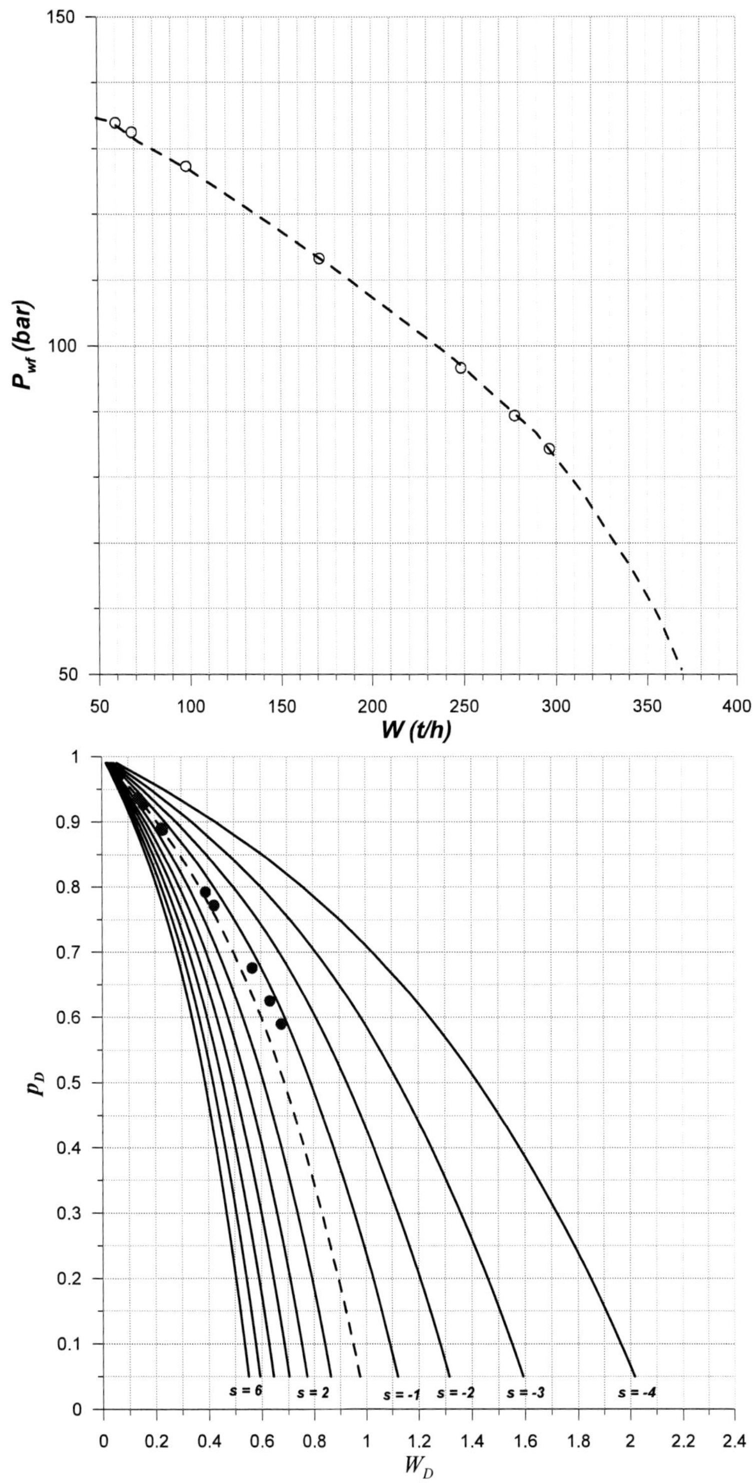

Figure 23. Continued on next page.

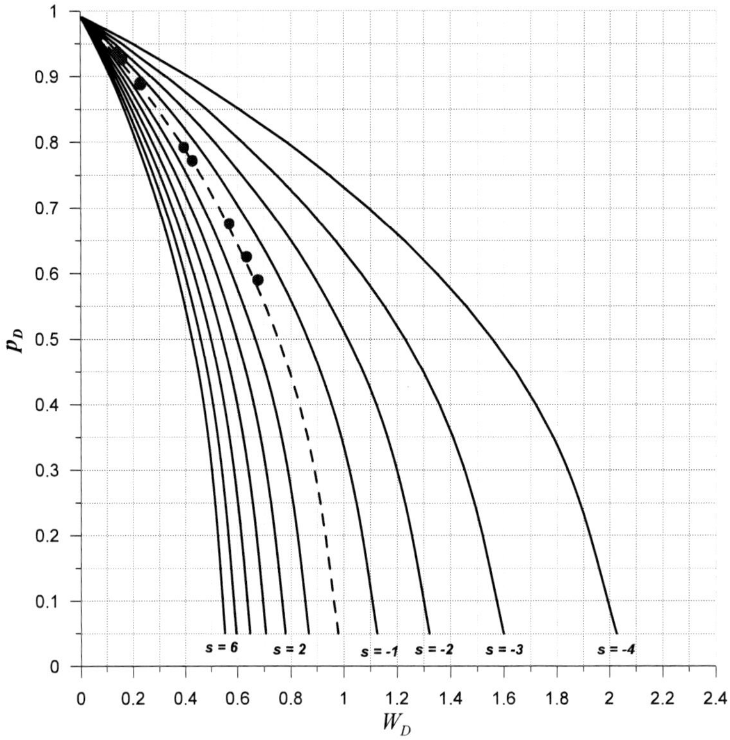

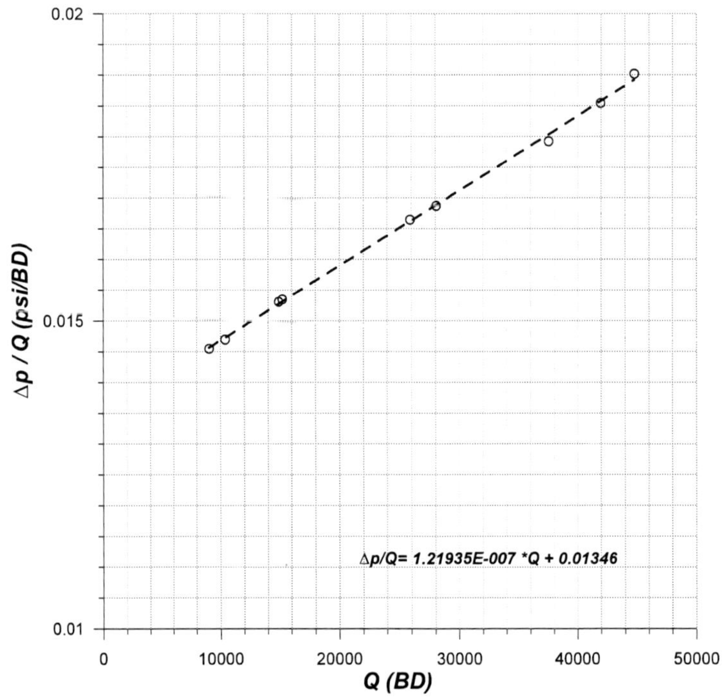

Figure 23. Overlapping of the inflow curve of well M-110 obtained in 1985 with the type- curves for low and high salinity. The diagnosed damage values s = -0.3 and s = -0.1 imply that there is no damage in the well. The ordinate b = 0.013, - - calculated with Jones' method, confirms the absence of damage, because b<0.05.

Well M-110 (Production Test of 1987)

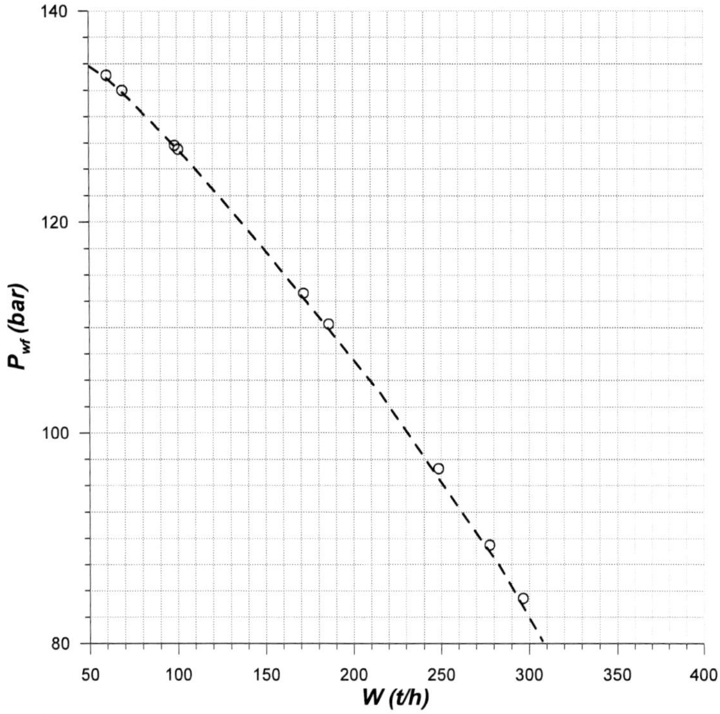

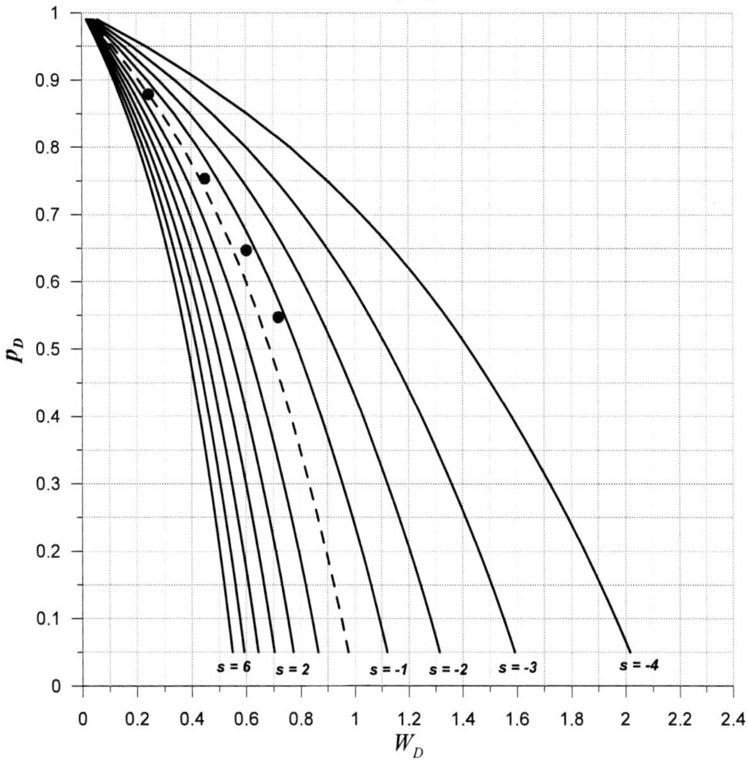

Figure 24. Continued on next page.

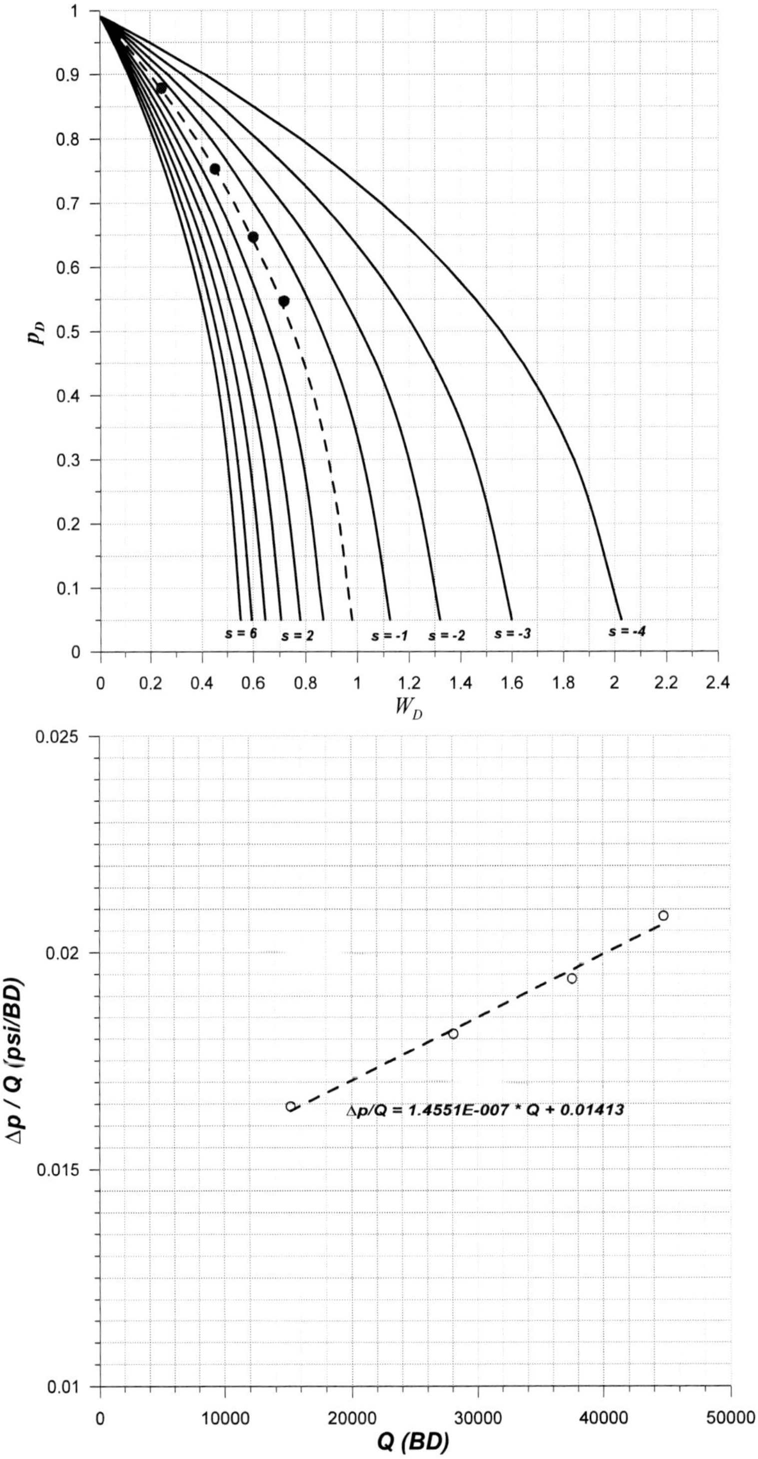

Figure 24. Overlapping of the inflow curve of well M-110 obtained in 1987 with the type- curves for low and high salinity. The diagnosed damage values s = -0.1 and - - s = 0 imply that there is no damage in the well. The ordinate b = 0.014, - - - - - calculated with Jones' method, confirms the absence of damage, because b< 0.05.

Discussion of Results

The type-curves with damage effect proposed in this work have demonstrated its effectiveness in the determination of damage in geothermal wells, from data of production tests.

In Table 1 we see a summary of the results obtained, by which the proposed methodology is validated. Jones' method allowed verifying the existence or absence of damage and, in all cases, there was agreement with the qualitative and quantitative diagnoses obtained with the proposed methodology.

Table 1. Validation of the proposed methodology by comparing the damage values obtained with Jones' qualitative method, using data of production tests in wells from three different fields.

Well	*S* damage value obtained with proposed methodology		Jones' method (1976)	
	Type-curve for low salinity	Type-curve for high salinity	b	Diagnosis
Carry City	0.4	0.6	0.063	Damage
Sweezy-1	-2.5	-2.2	0.036	No Damage
M-110 (initial conditions)	-0.5	-0.2	0.008	No Damage
M-110 (after 6 years)	-0.3	-0.1	0.013	No Damage
M-110 (after 8 years)	-0.1	0	0.014	No Damage

In Table 1 positive and negative values of s are presented according to the previous comment. Positive values of s indicate the presence of damage and negative values denote beneficial conditions for the well.

The value of the damage determined in each well corresponds to the stage of the well during its production test. This value changes with the time of exploitation and it also changes due to the manipulations carried out in the well (deepening, cleanings, stimulations, fracturing, etc.).

The behavior of the damage is a function of the characteristics of the reservoir and, for the same reason, is an indicative of its decline. This is proven in the behavior of the damage values obtained in well M-110 for its initial conditions, after 6 and 8 years of exploitation.

Since the proposed methodology is innovative and, for these tests, there is no field data available with which to verify the damage diagnosis, Jones' method was adopted; and even though it is only qualitative, it diagnoses correctly the presence or absence of damage. From the results of the presented validation we conclude that the proposed methodology, in combination with Jones' method, provides certainty in the diagnosis of damage in the well and its use is recommended.

Conclusions and Recommendations

From the results obtained in this research, the main conclusions are:

- A critical analysis of relevant bibliography was made; this included works from the beginning of reservoir engineering up until the most recent works in this field. The analysis was centered on the existent relationship between the skin factor and the inflow curves, with respect to the characterization of the system reservoir-well, in the feeding areas. The research covers oil and geothermal systems.
- As a result of this analysis we propose the first inflow type-curve with damage effect for geothermal reservoirs, with low and high salinity.
- This methodology is established to determine the value of damage in a well applying the inflow type-curves with damage effect. This methodology includes, as the last stage of Jones' qualitative method, the identification of the presence or absence of damage in a well and, in this way, to verify the obtained results.
- The methodology is innovative because previously the skin factor could only be determined from the analyses of transitory pressure tests. On the other hand, with the proposed methodology it is possible to determine the damage in a well from the measurements of its production parameters.
- The methodology was validated applying this technique to data from of five production tests of wells located in the United States of America and in the Cerro Prieto México, geothermal field.

The usefulness of the proposed methodology is manifested when monitoring the behavior of wells subjected to continuous exploitation and whose production cannot be suspended. In this way, the proposal of this research consists in disposing of at least three measurements; of flow, pressure and enthalpy, at different opening conditions, with which we are able to apply the diagnosis and evaluation methodology for the damage in the well caused by its continuous operation.

The use of instruments with acquisition data systems helps to eliminate reading errors of the parameters within the proposed methodology. With multiple readings, different regression methods can be used to identify if the abnormal responses correspond to the reservoir, or if they are related to transient effects or if they are only errors attributable to the instruments.

This research also involves new investigation lines. Even when the parameter M is of general application, it is convenient to analyze the characteristics of construction of the geothermal inflow relationships for low and high salinity (Eqs. 67 and 69). In this case a composition of gas in the mixture, equivalent to 5% in weight, was assumed. In accordance with this it is recommended to construct inflow curves for different percentages of gaseous fractions in the composition and to affect them with the damage effect.

Nomenclature

b	Value of the ordinate in Eqs. (32), (33) and (34)	
h	Reservoir thickness.	(ft)
m	Slope in Eqs. (32), (33) and (34)	
p	Pressure.	(psia)
q	Volumetric rate.	(cm^3/s)
r	Radius.	(ft)
s	Formation skin factor or damage effect	

B	Fluid volume factor of the formation.	(dimensionless)
H	Fluid enthalpy.	(kJ/kg)
F	Auxiliary variable in Fetkovich equation.	
J	Productivity index.	[(brl/dia)/psi]
K	Reservoir permeability.	(mD)
M	Variable of decline involving the damage effect.	
Pot.	Thermal power.	(MW)
Q	Volumetric rate.	(brl/d)
W	Mass rate.	(t/h)

Greek symbols

μ	Fluid viscosity.	(cp)
Δ	Difference.	

Subscripts

ag	Symbol related to water.
b	Fluid boiling condition.
D	Dimensionless.
e	Reservoir conditions.
f	Future conditions.
i	Initial conditions.
max	Maximum value of the variable.
o	Symbol related to oil.
w	Well conditions.
wf	Well bottom flowing conditions.

Superscripts

n	Decline factor in Eq. (9)
τ	Turbulence factor.

References

Agarwal, R. G., Al-Hussainy, R., Ramey, H. J. Jr. (1970). An investigation of wellbore storage and skin effect in unsteady liquid flow: I. Analytical treatment. *Soc. Pet. Eng., September*, pp. 279 – 290.

Al Qahtani, A. M. (2001). A new technique and field application for determining reservoir characteristics from well performance data. *SPE Middle East Oil Show and Conference, SPE 68141*, Bahrain, pp. 172 – 179.

Barua, J., Horne, R. (1997). Computarized analysis of thermal recovery well test data. *Society of Petroleum Engineers, Formation Evaluation,* pp. 560 – 566.

Chen, C. S., Chang, C. C. (2006). Theoretical evaluation of non-uniform skin effect on aquifer response under constant rate pumping. *Journal of Hydrology*, **317**, pp. 190 -201.

Chu, W. C., García-Rivera, J., Raghavan, R. (1980). Analysis of interference test data influenced by wellbore storage and skin effect at the flowing well. *Journal, Pet. Tech.*, **32** (1), p.171.

Chu, M. H. (1988). Inflow performance relationships for geopressured geothermal wells. *Geothermal Resources Council Transactions*, **12**, pp. 437 – 440.

Earlougher, R. C., Jr. (1977). *Advances in well test analysis*, Monograph Vol. 5, Dallas, TX.,U.S.A.: Society of Petroleum Engineers of AIME.

Evinger, H. H., Muskat, M. (1942). Calculation of theoretical productivity factor. *Trans., AIME*, No. **146**, pp. 126 – 139.

Fetkovich, J. J. (1973). The isochronal testing of oil wells. *SPE 4529*, presented at the *SPE 48th Annual Fall Meeting,* Las Vegas Nevada, U.S.A., pp. 78- 84.

Freeston, D. H., Hadgu, T. (1988). Comparison of results from some wellbore simulators using a data bank. *Proc. 10th New Zealand Geothermal Workshop*, Auckland, N. Z., pp. 408 – 414.

Gallice, F., Wiggins, M. (2004). A comparison of two-phase inflow performance relationships. Paper *SPE 88445, SPE Mid-Continent Operations Symposium*, Oklahoma City, U.S.A., pp. 100 - 104.

Garg, S. K. (1980). Pressure transient analysis for two-phase (water/steam) geothermal reservoirs. *Society of Petroleum Engineers Journal,* June, pp. 206 -212

Garg, S. K., Kassoy, D. R. (1981). *Convective heat and mass transfer in hydrothermal system in geothermal systems: Principles and case histories*, New York U.S.A.: Edited by L. Rybach and L. J. P. Muffler, John Wiley and Sons Ltd.

Garg, S. K., Riney, T. D. (1984). Analysis of flow data from the DOW/DOE L. R. Sweezy No. 1 well. Topical Report, *DOE/NV/10150-5*, La Jolla, CA. E.U.A., pp. 78 – 83.

Gilbert, W. E. (1954). *Flowing and gas-lift well performance*. Drilling and Production Pract., API, 126 p.

Goyal, K. P., Miller, C. W., Lippman, M. J. (1980). Effect of measured wellhead parameters and well scaling on the computed downhole conditions in Cerro Prieto Wells. *Proc. 6th Workshop on Geothermal Reservoir Engineering*, Stanford University, California, U.S.A. pp. 130- 138.

Grant, M. A., Donaldson, I. G., Bixley, P. F. (1982). *Geothermal reservoir engineering*, New York, U.S.A.: Academic Press.

Gunn, C., Freeston, D. (1991a). Applicability of geothermal inflow performance and quadratic drawdown relationships to wellbore output curve prediction. *Geothermal Resources Council Transactions,* **15**, pp. 471 – 475.

Hadgu, T., Zimmerman, R. W., Bodvarsson, G. (1995). Coupled reservoir-wellbore simulation of geothermal reservoir behavior. *Geothermics,* **24** (2), pp. 145 – 166.

Helmy, M. W., Wattenbarger, R. A. (1998). New shape factor for well produced at constant pressure. *SPE Gas Technology Symposium, SPE 39370*, Calgary Canada., pp. 532 – 538.

Horner, D. R. (1951). Pressure build-up in wells. *Proc. Third World Petroleum Congress, Section II*, E. J. Brill Leiden, p. 503.

Iglesias, E. R., Moya, S. L. (1990). Geothermal inflow Performance relationships. *Geothermal Resources Council Transactions,* **14** part II, pp. 1201 – 1205.

Iglesias, E. R., Moya, S. L. (1998). Applicability of geothermal inflow performance reference curves to CO_2-bearing reservoirs. *Geothermics,* **27** (3), pp 305 – 315.

James, R. (1968). Wairakey and Larderello: Geothermal power system compared. *N. Z. Journal Science*, **11** (4), pp. 706 - 715.

James, R. (1970). Factors controlling borehole performance. *Proc. United Nations Symp. On the development and use of geothermal resources, Geothermics, Sp. Iss.* No. **2**, pp. 1502 – 1515.

James, R. (1980). Deduction of the character of steam-water wells from the shape of the output curve. *Proc. N. Z. Workshop, Univ. of Auckland*, New Zealand., pp. 56 – 61.

James, R. (1983). Locus of wellhead pressure with time under production discharge. *Proc. N. Z. Workshop, Univ. of Auckland*, New Zealand., pp. 543. 549.

James, R. (1989). One curve fits all. *Proc. 14th Workshop on Geothermal Reservoir Engineering, Stanford University*, California, U.S.A., pp. 329 – 334.

Jones, L. G., Blount, E. M., Glaze, O. H. (1976). Use of short term multiple rate flow tests to predict performance of wells having turbulence. *SPE 51st Annual Fall Meeting, SPE 6133*, New Orleans, LA., U.S.A., pp. 378 – 383.

Kjaran, S. P., Elliasson, J. (1982). *Geothermal reservoir engineering:* Lecture notes, Universitty of Iceland, Reykjavik, Iceland.

Klins, M. A., Clark, L. (1993). An improved method to predict future IPR curves. *SPE Reservoir Engineering*, pp. 243 – 248.

Klins, M., Majcher, M. W. (1992). Inflow performance relationships for damaged or improved wells producing under solution-gas drive. *Journal Pet. Tech., SPE-AIME*, pp. 1357 – 1363.

Leaver, J. D., Freeston, D. H. (1987). Simplified prediction of output curves for steam wells. *Proc. 9th New Zealand Workshop, University of Auckland*, New Zealand, pp. 55 – 58.

Matthews, C. S., Russell, D. G. (1967). *Pressure buildup and flow test in wells*, Society of Petroleum Engineers of AIME, Monograph Vol. I, Dallas, Texas, U.S.A.: Henry L. Doherty Series.

Meza, C. O. (2005). *Efecto de la precipitación de sales en el diagnóstico de permeabilidades rocosas*, Tesis de maestría, Cuernavaca, Morelos, México: CENIDET, (Centro Nacional de Investigación y Desarrollo Tecnológico) SEP.

Millikan, C. V., Sidewell, C. V. (1931). Bottom-hole pressures in oil wells. *Trans. AIME*, pp. 194 – 205.

Montoya, D. (2003). *Estimación de permeabilidades de yacimientos geotérmicos mediante la aplicación de curvas tipo de influjo geotérmico*, Tesis de maestría, Cuernavaca, Morelos, México: CENIDET, (Centro Nacional de Investigación y Desarrollo Tecnológico) SEP.

Moya, S. L. (1994). *Efectos del bióxido de carbono sobre el transporte de masa y energía en yacimientos geotérmicos*, Tesis Doctoral, México: División de Estudios de Posgrado, Facultad de Ingeniería, Universidad Nacional Autónoma de México.

Moya, S. L., Iglesias E. R. (1995). Numerical simulation on carbon dioxide effects in geothermal reservoirs. *Proceedings of the TOUGH Workshop '95, Lawrence Berkeley Laboratory*, pp. 119 – 130.

Moya, S. L., Iglesias E. R., Aragón, A. A. (1995). Curvas de referencia adimensionales para estimar productividades de masa y energía de yacimientos geotérmicos con/sin bióxido de carbono. *Geotermia, Revista Mexicana de Geoenergía,* **11** (3), pp. 167 – 179.

Moya, S. L., Aragón, A. A., González, L. (1997). Estimación de curvas de producción de pozos geotérmicos empleando dos curvas de referencia adimensionales del comportamiento de influjo. *Ingeniería hidráulica en México*, **12** (3), pp. 35 – 40.

Moya, S. L., Aragón, A. A., Iglesias, E. R., Santoyo, E. (1998). Prediction of mass deliverability from a single wellhead measurement and geothermal inflow performance reference curves. *Geothermics,* **27** (3), pp. 317 – 329.

Moya, S. L., Uribe, D. (2000). Computational system to estimate formation permeabilities by superposition of the well inflow curve with geothermal inflow type curve. *Proceedings World Geothermal Congress*, pp. 2731 – 2737.

Moya, S. L., Uribe, D., Aragón A. A., Garcia, G. (2001). Formation permeability at the feedzone of geothermal wells employing type-curves. *Geofísica Internacional,* **40** (3), pp. 163 – 180.

Moya, S. L., Uribe, D., Montoya, D. (2003). Computational system to estimate formation permeabilities and output curves of geothermal wells. *Computers & Geosciences,* **29**, pp. 1071 – 1083.

Muskat, M. (1945). The production histories of oil producing gas-drive reservoirs. *Journal Applied Physics,* **16**, pp. 147 – 153.

Pruess, K., Shroeder, R. C. (1980). *SHAFT79 Users's manual*, Report LBL-10861, Berkeley Cal. U. S. A.: Lawrence Berkeley Laboratory.

Pruess, K. (1987). *TOUGH User's guide*, Report LBL-20700, Berkeley, Cal. U.S.A.: Lawrence Berkeley Laboratory.

Pruess, K., Oldenburg, C., Moridis, G. (1999). *TOUGH2 User's guide, Version 2.0*, Report LBNL-43134, Berkeley Cal., U.S.A.: Lawrence Berkeley Laboratory.

Ramey, H. J., Jr. (1970). Short-time well test data interpretation in the presence of skin effect and wellbore storage. *Journal Pet. Tech.*, pp. 97 – 104.

Ramey, H. J., Jr. (1981). *Reservoir engineering assessment of geothermal system*, Cal. U.S.A.: Notes of Stanford University.

Ribó, M. O. (1989). Análisis de pruebas de presión en pozos de Cerro Prieto. *Proceedings Symposium in the field of geothermal energy, Convenio entre Comisión Federal de Electricidad y el Departamento de Energía de los Estados Unidos de Norteamérica*, San Diego California, Estados Unidos de Norteamérica, pp. 123 – 129.

Rivera, J. R., Ramey, H. J. Jr. (1977). Application of two-rate flow tests to the determination of geothermal reservoirs parameters. *SPE 6887, Proc. 52th Annual Fall Meeting of Society of Petroleum Engineers*, Denver Col. U.S.A., pp. 346 – 353.

Standing, M. B. (1970). Inflow performance relationships for damaged wells producing by solution-gas drive. *Journal Pet. Tech.*, pp. 1399 – 1400.

Van Everdingen, A. F., Hurst, W. (1949). The application of the LaPlace transformation to flow problems in reservoirs. *Petroleum Transactions AIME,* **196**, pp. 156 – 164.

Verma, S., Días-González, L., Sánchez-Upton, P., Santoyo, E. (2006). OYNYL: A new computer program for ordinary, York and New York least-squares linear regressions. *WSEAS Conference Proceedings on EED´06*, Venice, Italy, in press, 6 p.

Vogel, J. V. (1968). Inflow performance relationships for solution gas drive wells. *Journal Pet. Tech. SPE 1476 Annual Fall Meeting of Society of Petroleum Engineers*, Dallas Texas, U.S.A., pp. 66 – 79.

Warren, J. E., Root, O. J. (1963). The behavior of naturally fractured reservoirs. *Society of Pet. Eng.*, pp. 245 – 255.

Weller, W. T. (1966). Reservoir performance during two-phase flow. *Journal Pet. Tech.*, pp. 240 – 246.

Wiggins, M. L. (1994). Generalized inflow performance relationships for three-phase flow. *SPE Production Operations Symposium SPE 25458*, Oklahoma City, U.S.A. pp. 275 – 286.

INDEX

D

E

F

G

H

I

T

U

V

W

X

Y